POWER CYCLES
and ENERGY EFFICIENCY

POWER CYCLES
and ENERGY
EFFICIENCY

E.J. Hoffman

ACADEMIC PRESS

San Diego   New York   Boston   London   Sydney   Tokyo   Toronto

Copyright © 1996 by ACADEMIC PRESS

Academic Press, Inc.
525 B Street, Suite 1900, San Diego, California 92101-4495, USA
http://www.apnet.com

Academic Press Limited
24-28 Oval Road, London NW1 7DX, UK
http://www.hbuk.co.uk/ap/

Library of Congress Cataloging-in-Publication Data

Power cycles and energy efficiency / [edited] by E. J. Hoffman.
        p.   cm.
    Includes index.
    ISBN 0-12-351940-3 (alk. paper)
    1. Thermodynamics.   2. Power (Mechanics)   3. Energy conservation.
    I. Hoffman, E. J. (Edward Jack), date.
    TJ265.P69   1996
    621.402'1--dc20   96-13577
                                                                   CIP

PRINTED IN THE UNITED STATES OF AMERICA
96 97 98 99 00 01 QW 9 8 7 6 5 4 3 2 1

Contents

CHAPTER 2
The Conversion between Heat and Pressure – Volume Work   45

Preface

Understanding the conversion of heat to work—and sometimes the opposite—is the foremost objective of thermodynamic investigations. The former is manifested by power cycles of one sort or another, the latter by the heat pump or refrigeration cycle. Moreover, there may be variations in which compositional or chemical changes are involved, such as in absorptive refrigeration. Another variation of interest is the latent heat pump, including the special case of vapor recompression.

The embodiment of principal interest is that of a nonreactive working fluid, where combustive support is used to furnish the heat to the working fluid. Gaseous-phase expansion of the working fluid produces pressure–volume work, and in turn mechanical work, which may be transformed to electrical energy as in central electric power generating stations. Of less interest here and noted mainly in passing are cycles involving internal heat generation, whereby the working fluid itself consists of combustants and combustion products.

These energy conversion cycles or processes will involve efficiency losses, of varying degree, which may be limiting. Thus the classic Rankine cycle, practical as it is, has limitations from the standpoint of efficiency, for which there appears to be no remedy other than operating under as severe a set of conditions as is feasible.

Even the hypothetical Carnot cycle, employed as the reference standard for establishing the highest theoretical efficiencies attainable, has theoretical inadequacies. As shown in the pages to follow, this cycle is not necessarily the last word.

The remaining options involve the Joule cycle, which is closed, or the Brayton cycle, which is open. (These descriptors are sometimes used interchangeably; e.g., the Brayton cycle is an open Joule cycle, and the Joule cycle is a closed Brayton cycle.) Originally called the air engine or gas engine, the ideal Joule cycle consists of the addition of heat to a gaseous working fluid, followed by an adiabatic expansion, heat rejection, and then recompression of the working fluid to complete the cycle. On the other hand, the Brayton cycle is that of the gas-fired/turbine generator, which may or may not use regenerative heat transfer between the combustion products and the combustants or reactants.

In the Joule cycle as well, however, there are limitations. One limitation is the temperature at which heat must be rejected to the surroundings. Another limitation is that the fuel efficiency is generally low, since the combustive support (or heat added) must occur at relatively high temperatures—that is, the combustion products must leave the system at a relatively high temperature.

However, there are variations that can be deployed to get around these difficulties to a greater or lesser extent to enhance the overall efficiency. This aspect will be of paramount concern. Of particular interest here is the use of interstage heat addition and rejection utilizing the Joule cycle or a variation thereof, thereby controlling the path for compression and expansion. In this case, both the compression and expansion sections are non-adiabatic.

Such a cycle has great potential in principle, although the hardware requirements and heat transfer requirements may be limiting to a degree. The applications may be tied to a number of heat and thermal energy sources. These applications include the combustion of fossil fuels and the utilization of chemical and nuclear reactions. Additionally, there is the use of both high-grade and low-grade thermal sources. Examples of the former include high-temperature, high-pressure water, high-temperature steam, and high-temperature gases such as may be produced from nuclear reactors or fossil fuel steam generators or boilers. Examples of the latter include waste heat from chemical and petroleum processing, and notably from power generation and geothermal energy. Not least is the potential for using solar energy in this manner.

Of interest also is the use of the cycle described above in cogeneration or combined power cycle operations. While it can be used first as a topping cycle, its greater promise is as a bottoming cycle in which the low-grade heat produced from the primary cycle is used.

Of additional interest is the use of the reverse Joule cycle and the reverse Rankine cycle, not only for a heat pump but also for a latent heat pump. This, too, will be addressed in detail, especially for drying, and will include means of enhancing the performance efficiency or coefficient of performance.

As an introduction to the subject, the energy balance will be carefully reviewed and examined with respect to enthalpic or heat energy changes and the conversion of heat to work, and to the inclusion versus the exclusion of dissipative effects. This is a prominent feature of the relationships for fluid flow and is basic to the thermodynamics of power cycles.

The Carnot and Rankine cycles are subsequently examined and compared along nontraditional lines, as are cycles involving nonadiabatic compression and expansion. (Even the idea of entropy is analyzed as to its deviation from exactness. In general, entropy is not a state function, unlike enthalpy.) Under certain conditions, in fact, the Rankine efficiency can be perceived as greater than the Carnot efficiency, although the comparison is something like apples and oranges since the cycles are intrinsically different. There is an indication that the Carnot cycle should not always be viewed as the limiting case. In other words, there are ways around it.

The fact that anomalies to the conventional wisdom may indeed occur can serve to revitalize the subject of energy cycles. The demonstration that at least one energy or power cycle circumvents the perceived restrictions of the Carnot cycle indicates the possibility that others may exist. In sum, there is the intriguing prospect of enhanced efficiencies for the conversion of heat to work to electricity.

E. J. Hoffman
Laramie, Wyoming

CHAPTER I

Energy
Relationships

The fundamental premise that energy is conserved provides the rationale upon which the study of thermodynamics rests. This conservative property is, in fact, inherent to the very nature of energy: The concept arises as a constant in the mathematical operations of integrating or solving the differential equations that describe the behavior of a system. This property of constancy, called the first law of thermodynamics, permits the system behavior to be phrased in terms of an overall or total energy balance: It is the base upon which the mathematical relationships and calculations find a firm footing.

In the dynamics of motion for rigid bodies, the variables that describe system behavior are position and time, which may alternately be encoded as position and the positional derivatives or velocity components. The integration or solution of Kepler's laws of planetary motion leads to the relationship that expresses total energy as a sum of kinetic and potential energy terms, and the more familiar expression for total energy change as the sum of the kinetic energy change and potential energy change. The total energy change may be expressed in its integrated form for discrete or incremental changes, or in its differential form for an infinitesimal change. [These and other aspects of the abstract concept of energy are dealt with at great length in another of the author's books (1).]

In thermodynamics, which involves the behavior of gaseous and liquid systems, the additional independent variables of temperature and pressure are included. (For open or flow systems, volume or specific volume is considered a dependent variable by virtue of the equation of state for the gas or liquid. Alternatively, for closed systems, temperature and volume can be regarded as the independent variables and pressure can be regarded as the dependent variable.) Composition may be viewed as a parameter for nonreacting systems or, in the case of chemically reacting systems, as a variable.

The inclusion of temperature and pressure as variables leads to the concept of the heat function or enthalpy function for flow or open systems, and to the concept of internal energy for closed systems. The concept of enthalpy first arises as a constant of integration in solving or integrating the Joule–Thomson behavior for a gas or liquid; that is, the Joule–Thomson behavior expresses the differential change in temperature with respect to a differential change in pressure for a gas (or liquid) flowing

1

through a valve or porous plug. (As near as possible, Joule–Thomson determinations are conducted as an isolated flow system, with no energy exchange with the surroundings. The system is therefore referred to as "isenthalpic.") The resulting derivative, that is, the differential change in temperature with pressure, is called the Joule–Thomson coefficient. The Joule–Thomson coefficient $\mu = dT/dP$ is in turn a function of temperature and pressure. The differential equation so obtained may, in principle, be solved or integrated. The mathematical operation introduces a constant of integration which may be generalized to a function called the heat function or more commonly the enthalpy or enthalpy function.

A closed system can be analyzed in similar fashion. The so-called Gay–Lussac behavior, which is the change in temperature with respect to volume for an expanding gas (or liquid) in a closed system, may be represented similarly in terms of temperature and volume, and the mathematical operation or solution introduces a constant of integration which may be generalized to a function called the internal energy or internal energy function. Furthermore, the internal energy and enthalpy are mathematically related.

The idea of temperature and a temperature scale is of primary importance, and it turns out that the creation or calibration of an absolute or thermodynamic temperature scale is, in actuality, a correlation made to establish and preserve basic thermodynamic relationships, namely, the perfect or ideal gas law and the Clausius–Clapeyron relation for latent heat change. Further minor adjustments can be made to the thermodynamic scale to more accurately accommodate the derived relationship for the Joule–Thomson coefficient or Gay–Lussac coefficient. (The subject is amplified in reference 1.)

The fundamental notion of pressure also requires further clarification. Pressure is a measurement related to the position of a mass or weight in a gravitational field, in this case the Earth's gravitational field. That is, the system is connected to a "dead" weight which provides a measure of what we call pressure. Changes in pressure can be indicated by a change in mass of the dead weight. Calibration is customarily made to a pressure gauge.

This feature of pressure being measured in terms of the mass and position of a dead weight in the Earth's gravitational field connects pressure to potential energy and potential energy change, and gives rise to the idea of pressure–volume work. In other words, the change in position of the dead weight in the Earth's gravitational field with a change in pressure is translated as work in the most fundamental sense. Thus a change in pressure is manifested as work, and in the foregoing context it is referred to as pressure–volume work in that it will have the dimensions of pressure times volume. If pressure is measured as the mass of the dead weight per unit area, then pressure times volume will be expressed in units of mass-distance—the customary units for work.

For flow or open systems, as will be shown, the pressure–volume work term will be volume times a change in pressure. For a closed system it will be the negative of pressure times a change in volume. The terms may be interconnected whereby the difference between enthalpy and internal energy is pressure times volume.

The corresponding pressure–volume work term may be incorporated into the enthalpy function or into the internal energy function. As it turns out, another term is required to make up the difference. For the purposes herein, this term is called the intrinsic energy or intrinsic energy change. It also may be called the "lost work," and for other purposes may be made equal to the absolute or thermodynamic temperature multiplied by a change in the entropy function—a matter to be further discussed else-where. Basic to the inclusion of intrinsic energy is the idea of irreversibil-ity, in that the intrinsic energy change or lost work is always positive in the direction of flow or change; that is, it always acts against system changes. Accordingly, the manifestations are often called frictional or dissipative effects, or irreversibilities.

The heat function or enthalpy function is basic to the idea of heat or heat energy. A generalized function of temperature and pressure, by virtue of its derivation, enthalpy is a state function of temperature and pressure; that is to say, it is a function of the point values of temperature and pressure, and changes in its value are independent of any relationship which might exist between temperature and pressure. (Stated another way, its differential is said to be "exact.") In its complete form, a change in enthalpy occurs from both a change in temperature and a change in pressure, although the change in temperature is generally controlling.

These enthalpic changes may in turn be related to pressure–volume work and the intrinsic energy change. For purposes of simplification, the latter may sometimes be neglected, as in the "isentropic" compression or expansion of a gas.

Similar remarks may be made for the internal energy function.

These things said, there is the matter of relating the changes that occur in a fluid, which may be a gas or liquid, to an energy exchange with the surroundings; that is, between the heat added to the fluid and the work done by the fluid. For a flow system, the net change between the heat added and the work done will be manifested as a change in the kinetic and potential energy terms plus a change in the enthalpy function. This change is known as the total energy balance. (A similar relationship may be set forth for internal energy changes alone, since in this case kinetic and potential energy changes are not a factor.)

The energy terms that are to be included or neglected in the overall energy balance will depend upon the particular system configuration under consideration, which most usually will be considered as a flow or open system. In net sum, a way to interconnect velocity or kinetic energy effects, potential energy effects, and enthalpic effects is afforded. The enthalpic effects in turn may be expressed either as pressure–volume work effects plus dissipative effects, or as heat or thermal energy effects in terms of temperature and pressure.

Thus the total or overall energy balance provides the means by which heat and work exchange can be related variously to kinetic and potential energy changes, to pressure–volume work, and to enthalpic energy change. Enthalpic energy change can then be related to the pressure–volume work change plus dissipative changes or irreversibilities, or, alternatively, can be related to temperature and pressure changes via the heat capacity and Joule–Thomson coefficient. Furthermore, and importantly, pressure–volume work change and dissipative effects can be related to temperature and pressure changes in this way.

The connection between these two alternative means of representation and its simplifications will generate the classic expressions for the temperature–pressure behavior (or temperature–volume behavior) for an idealized isentropic compression or expansion—a basic feature of power cycles. The comparison between heat added and work done is the criterion by which power cycles are evaluated in terms of efficiency.

The subject is further developed as follows: we begin with the twin concepts of pressure and pressure–volume work, followed by the connection of pressure–volume work to enthalpy and internal energy, and the origins and derivations for the latter energy forms. The elemental relationships and correlations for fluid flow are presented as a matter of course and both adiabatic isentropic behavior and nonadiabatic behavior are compared. Further refinements in some of the relationships for fluid flow complete the chapter. This introductory material is followed in the succeeding chapters by an emphasis on the conversion of heat to work—the raison d'être of thermodynamics. The accent is on increased efficiencies for energy conversion cycles.

1.1. THE PRESSURE SCALE AND PRESSURE–VOLUME WORK

For the total energy of a flow system, in terms of kinetic energy (K.E.) and the potential energy (P.E.), the differential energy change is

$$dE' = d(\text{K.E.}) + d(\text{P.E.}) = \tfrac{1}{2}m\,dv^2 + mg\,dz,$$

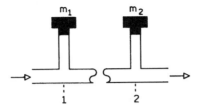

Fig. 1.1. Flow system using dead-weight testers to measure pressure and pressure drop.

where

$$E' = \text{energy in force–distance units,}$$
$$m = \text{mass of flowing system, or mass–time,}$$
$$v = \text{velocity,}$$
$$z = \text{elevation,}$$
$$g = \text{acceleration of gravity.}$$

The potential energy term P.E. directly converts to pressure and pressure–volume work.

To calibrate the pressure scale to mass using a dead-weight tester, consider first the flow system shown in Fig. 1.1. Thus

$$\frac{m_1 g}{A} = P_1', \qquad \frac{m_2 g}{A} = P_2',$$

where A is the cross-sectional area of each fluid column and P' denotes pressure in force–distance units. The quantity m is the mass of the dead weight.

For a column of fluid of height Δz, the difference in potential energy represented by the two dead weights is

$$\Delta(\text{P.E.}) = (m_2 - m_1) g \, \Delta z = A \, \Delta z (P_2' - P_1') = V(P_2' - P_1').$$

For a differential change,

$$d(\text{P.E.}) = g \, \Delta z \, dm = V \, dP',$$

where V is arbitrary, as is Δz. Referenced to sea level,

$$\frac{1}{g_c} d(\text{P.E.}) = V \, dP,$$

where P is in mass–distance units and $P' = g_c P$. The value g_c is the acceleration of gravity measured at sea level, and it becomes the conversion constant between force–distance and mass–distance units.

(*Note*: By the conventions used, force equals mass times acceleration. Accordingly, force divided by mass equals acceleration or force–distance divided by mass–distance equals acceleration, which is represented by g or by g_c at sea level.)

The term $V\,dP$ denotes the pressure–volume or volume–pressure work change for a flow system.

Closed System

For a closed fluid system, the calibration of pressure in force–distance units is given by

$$\frac{mg}{A} = P',$$

whereby the fluid is enclosed in a vertical cylinder of cross-sectional area A connected to a dead weight. If the height changes by dz, then

$$d(\text{P.E.}) = mg\,dz = P'A\,dz$$

or, in mass–distance units,

$$\frac{1}{g_c}d(\text{P.E.}) = m\frac{g}{g_c}\,dz = PA\,dz = P\,dV.$$

Corrected to sea level, $g = g_c$.

The term $P\,dV$ represents the pressure–volume work change, and the sign used will depend upon the conventions of the energy balance.

1.2. ENTHALPIC ENERGY CHANGE AND PRESSURE–VOLUME WORK

For flow or open systems, the enthalpic energy change is representable by (2, 3)

$$dH = \frac{\partial H}{\partial T}\,dT + \frac{\partial H}{\partial P}\,dP + C_p\,dT + \frac{\partial H}{\partial P}\,dP,$$

where

$$H = \text{enthalpy function,}$$

$$T = \text{temperature (absolute),}$$

$$P = \text{pressure,}$$

$$C_p = \text{heat capacity.}$$

For an isenthalpic or Joule–Thomson expansion,

$$dH = 0 = C_p\, dT + \frac{\partial H}{\partial P}\, dP$$

or

$$\frac{dT}{dP} = \mu = \frac{-1}{C_p} \frac{\partial H}{\partial P} \quad \text{and} \quad \frac{\partial H}{\partial P} = -C_p \mu = \phi \text{ or } \phi_T,$$

where μ is the Joule–Thomson coefficient and ϕ or ϕ_T is called the isothermal Joule–Thomson coefficient. Accordingly, in heat units,

$$dH = C_p\, dT - C_p \mu\, dP.$$

Neglecting kinetic and potential energy effects, the sign convention to be used is such that, per unit mass or mole of flow,

$$dq - dw = J_0\, dH,$$

where

$$q = \text{heat added to system,}$$

$$w = \text{work done by system,}$$

$$J_0 = \text{mechanical equivalent of heat.}$$

In an adiabatic expansion through an expander, for instance, where $dq = 0$, dw will be positive, whereas dH and dP will be negative. That is, mechanical work will be performed by the fluid system. Also, for a positive value of the Joule–Thomson coefficient, if dP is negative, dT will be negative.

In turn, the pressure–volume work change within the fluid can be incorporated into the enthalpy change,

$$J_0\, dH = V\, dP + d\omega \quad \text{or} \quad J_0\, dH - V\, dP = d\omega,$$

where, in consistent units,

$$V = \text{specific volume,}$$

$$\omega = \text{intrinsic energy in mechanical units.}$$

The intrinsic energy change $d\omega$ is defined by the difference. Also

$$d\omega = J_0 \, d\Omega,$$

where $d\Omega$ is the intrinsic energy change in heat units. Furthermore, in mechanical units,

$$d\omega = d\text{lw} = J_0 T \, dS,$$

where

$$d\text{lw} = \text{lost work change,}$$

$$dS = \text{entropy change.}$$

It is the usual convention to eliminate J_0 with the understanding that all units are to be consistent. For example, in principle, $R = J_0 R_0$, where the gas constant R is in mechanical units and R_0 is in heat units. Most often, however, R_0 is written simply as R, with the implication that it is in heat units. In other words, it is defined by its usage.

It is also customary use a bar above the various extensive variables, namely, V and the heat and work quantities, q, w, and lw, to denote that they are on a unit mass or unit mole basis. For the purposes of simplicity, this convention will be dropped here, although it is to be understood.

Therefore, there is the equivalence that, in consistent units,

$$dH = C_p \, dT - C_p \mu \, dP = V \, dP + d\omega.$$

If $d\omega = d\text{lw} = T \, dS = 0$, then the enthalpic change can be described as isentropic. Furthermore, for a perfect gas it may be assumed that the Joule–Thomson coefficient is zero; that is, $\mu = 0$.

Either of the foregoing enthalpic expressions may be incorporated into the total energy balance. Thus, for instance, on a unit mass basis,

$$dq - dw = \frac{1}{2g_c} \, dv^2 + \frac{g}{g_c} \, dz + V \, dP + d\text{lw}.$$

By eliminating one or another of the terms, several different circumstances can be accommodated, as will be demonstrated subsequently.

It may be noted that the term $d\text{lw}$ can pertain either to an isolated system, where $dq - dw = 0$, or to an isothermal system, where the difference will exist, since heat must be added or removed to maintain a constant temperature; in other words, dq or its derivatives will not equal zero. Therefore, we may write

$$d\text{lw}_{\text{isolated}} - dq = d\text{lw}_{\text{isothermal}}$$

and

$$-dw = \frac{1}{2g_c}dv^2 + \frac{g}{g_c}\,dz + V\,dP + d\text{lw}_{\text{isothermal}}.$$

For most purposes it is understood that the symbol $d$lw stands for $d\text{lw}_{\text{isothermal}}$, and it is so correlated.

1.3. INTERNAL ENERGY AND PRESSURE–VOLUME WORK

For a closed system, the internal energy change is representable by (2, 3)

$$dU = \frac{dU}{dT}\,dT + \frac{\partial U}{\partial V}\,dV + C_v\,dT + \frac{\partial U}{\partial V}\,dV,$$

where

$$U = \text{internal energy function,}$$
$$C_v = \text{heat capacity at constant volume.}$$

For a Gay–Lussac expansion (the expansion of an isolated closed system),

$$dU = 0 = C_v\,dT + \frac{\partial U}{\partial V}\,dV$$

or

$$\frac{dT}{dV} = \eta = -\frac{1}{Cv}\frac{\partial U}{\partial V} \quad \text{and} \quad \frac{\partial U}{\partial V} = -C_v\eta = \lambda \text{ or } \lambda_T,$$

where η is the Gay–Lussac coefficient λ or λ_T is called the isothermal Gay–Lussac coefficient. Accordingly,

$$dU = C_v\,dT - C_v\eta\,dV.$$

The sign convention to be used is again such that

$$dq - dw = J_0\,dU.$$

For an adiabatic mechanical expansion, $dq = 0$ and dw will be positive, whereas dU is negative and dV is positive. For a negative value of the Gay–Lussac coefficient, dT would be negative. It is understood that V is the specific volume—the volume per unit mass or mole.

Consequently, the internal energy change can be written as

$$J_0\, dU = -P\, dV + d\omega,$$

where $d\omega$ is the intrinsic energy change in mechanical units. It may or may not be equal to the intrinsic energy change for an open system, but, for reasons of consistency, it is assumed to be. Furthermore, $d\omega = d\text{lw} = T\, dS$, as for an open or flow system.

Since $P\, dV = d(PV) - V\, dP$, then in consistent units, dropping the mechanical equivalent of heat in the notation, if

$$dU = -P\, dV + d\omega \quad \text{and} \quad dH = V\, dP + d\omega$$

and if $d\omega$ is the same,

$$dU = d(PV) - dH \quad \text{or} \quad dH = dU + d(PV).$$

This is the customary assigned relationship between the internal energy U for a closed system and the enthalpy H for a flow system.

1.4. JOULE–THOMSON AND GAY–LUSSAC EXPANSIONS FOR A PERFECT GAS

For a perfect gas it is customary to assume that $\mu = 0$ and $\eta = 0$. This assumption is affirmed as follows.

As previously stated,

$$\begin{aligned}
d(PV) &= dH - dU \\
&= C_p\, dT - C_p \mu\, dP - (C_v\, dT - C_v \eta\, dV) \\
&= (C_p - C_v)\, dT - C_p \mu\, dP + C_v \eta\, dV \\
&= R\, dT - C_p \mu\, dP + C_v \eta\, dV
\end{aligned}$$

or

$$0 = -C_p \mu + C_v \eta \frac{dV}{dP}.$$

Since the derivative dV/dP will exist, it is a requirement that $\mu = 0$ and $\eta = 0$.

It may be observed also that if it is assumed that $P\,dV + V\,dP = R\,dT$, then, as is apparent,

$$\frac{dP}{dT} = \frac{R - P(dV/dT)}{V}$$

or

$$\frac{dT}{dP} = \frac{V(dT/dV)}{R - (dT/dV) - P}.$$

Consequently,

$$\mu = \frac{V\eta}{R\eta - P} \quad \text{or} \quad \eta = \frac{P\mu}{R\mu - V}.$$

From the preceding statements, if $\mu = 0$, then $\eta = 0$ or vice versa.

We may conclude, therefore, that the temperature does not change during the Joule–Thomson or Gay–Lussac expansion of a perfect gas. We also may conclude that the respective heat capacities are a function of temperature only, such that $dH = C_p\,dT$ and $dU = C_v\,dT$; that is, the heat capacity becomes independent of pressure or volume.

1.5. THE ENERGY BALANCE AND FRICTIONAL EFFECTS IN FLUID FLOW

The energy balance serves perhaps its most ubiquitous purpose in applications to single-phase flow in circular ducts, i.e., pipes or tubing. Full advantage may be taken of elevation changes, velocity changes, and pressure–volume changes, including dissipative effects. Extension to two-phase flow may be made by reference to single-phase flow. The energy balance also is involved, in one way or another, in orifice, venturi, pitot, and nozzle relationships and in the calculations for hydraulic pumps and turbines and for gas-phase compressors and expanders.

Frictional losses or dissipative effects in fluid flow involve the intrinsic energy change or lost work term. This term may be treated as a separate entity (as for the flow of fluids within a confining surface or surfaces) or incorporated into a coefficient or multiplier (as in the case of the discharge coefficient for the calibration and measurement of flow rate). Moreover, flow may be in either the viscous or turbulent region or in the transition zone in between. The subject is allied with boundary layer analysis, which will not be developed here, and extends to the treatment of heat and mass transfer rates, in being related to boundary layer thickness.

As a general rule, in fluid flow per se there will not be a work exchange with the surroundings. Moreover, the flow may be regarded as isolated; that is, adiabatic without a work exchange. Alternatively, the flow system may be regarded as nonadiabatic with heat exchange with the surroundings. A special case of the latter is isothermal flow or near-isothermal flow.

If a work exchange occurs, the subject is entered of the pumping of liquids, as well as the compression of gases—or the inverse, the hydraulic-driven turbine, and the gas expander or turbine. The flow embodiments pertain to horizontal and vertical flow and to the subject of particle motion relative to a fluid or flow flow relative to a particle. These subjects and related matters are taken up in this section.

Correlations for Horizontal Flow

Correlation is generally made to some form of the Fanning friction factor relationship by using horizontal steady-state flow with no work exchange. Accordingly,

$$-V\,dP = d\mathrm{lw} = f\frac{v^2}{2g_c D}\,dL,$$

where the term on the right is the friction factor relationship and where

$$f = \text{friction factor,}$$
$$D = \text{diameter or other dimension,}$$
$$L = \text{length.}$$

For flow through pipes or tubing, D becomes the inside diameter. It is understood that all units are consistent.

Correlation is made to the dimensionless Reynolds number Re,

$$\mathrm{Re} = \frac{Dv\rho}{\mu},$$

where ρ is the density and μ is the viscosity of the fluid, in consistent units. A correlation is shown in Fig. 1.2 for f against Re for parameters of relative internal surface roughness (2, 3).

For flow through packed beds or porous media, correlation may be made to

$$-V\,dP = d\mathrm{lw} = f\frac{v^2}{2g_c D_p}\,dL,$$

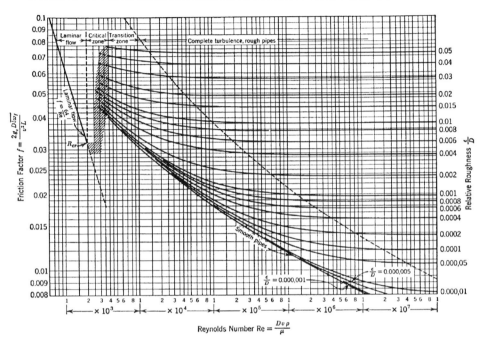

Fig. 1.2. Friction factor correlation for flow through pipes. [After L. F. Moody, *Trans. ASME 68*, 672 (1944). Redrawn in G. G. Brown *et al.*, *Unit Operations*, Wiley, New York, 1950.]

where D_p is the particle diameter or equivalent particle diameter. The correlation is made against the particle Reynolds number $\mathrm{Re} = D_p v \rho / \mu$ (3).

Other notations also can be used for the friction factor. Thus the substitutions that $f = 2f' = 4f''$ can be made. The correlating relationships will be affected by whichever substitution is made.

Viscous Flow. In well-behaved flow or viscous flow, the friction factor can be represented by

$$f = \frac{C}{\mathrm{Re}},$$

where C is a constant that depends upon the flow system and flow geometry. Thus

$$-V\,dP = -\frac{1}{\rho}\,dP = \frac{C\mu}{Dv\rho}\,\frac{v^2}{2g_c D}\,dL,$$

from which

$$v = \frac{2g_c}{C}\frac{D^2}{\mu}(-)\frac{dP}{dL} = \frac{K}{\mu}(-)\frac{dP}{dL}.$$

The permeability K is defined by the substitution. The preceding statement is equivalent to Poiseuille's law for flow through pipes, tubing, or capillaries. Alternately, it represents Darcy's law (d'Arcy) for flow through porous media.

For particle flow or flow around particles, the definition for viscosity yields Stokes' law for flow relative to spheres. Substituted into a force balance, the correlating relationship is obtained (3, 4). For other flow geometries, the viscous relationships are derived in reference 4.

Vertical Flow

The energy balance per unit mass flowing is denoted by

$$d\bar{q} - d\bar{w} = \frac{1}{2g_c}dv^2 + \frac{g}{g_c}dz + \bar{V}dP + d\overline{lw},$$

where, in consistent units, $dlw = d\omega = T\,dS$; that is, the change in lost work is also the change in intrinsic energy or the change $T\,dS$. Using the friction factor form for isothermal flow, and with no work transfer,

$$0 = \frac{1}{2g_c}dv^2 + \frac{g}{g_c}dz_{elev} + \bar{V}dP + f\frac{v^2}{2g_c}dL.$$

In upflow, $dz_{elev} = dL$. Using the compressibility factor form for the specific volume $\bar{V}$, where ρ is the density and where here z is the compressibility factor,

$$\bar{V} = \frac{1}{\rho} = \frac{zRT}{MP}.$$

Unfortunately, the compressibility factor z and the elevation z share the same symbol in the terminologies generally used. Confusion is avoided in the particular instance by the requirement that $dz = dz_{elev} = dL$. Therefore,

$$0 = \frac{1}{2g_c}dv^2 + \frac{zRT}{MP}dP + \left[\frac{g}{g_c} + f\frac{v^2}{2g_cD}\right]dL,$$

where the friction factor f is a function of the Reynolds number as indicated in Fig. 1.2 and may be designated symbolically as

$$f = f(\text{Re}) = f\frac{Dv\rho}{\mu},$$

where here μ is the viscosity.

For flow in a pipe at a constant mass velocity G,

$$v = \frac{W}{A\rho} = \frac{W}{A}\frac{zRT}{MP} = G\frac{zRT}{MP},$$

where W is the mass flow rate, A is the cross-sectional area, and G is the mass velocity. The mass flow rate W by definition is constant for steady-state flow. If A (or the diameter D) is assumed constant, then G will be constant. Further substituting the preceding equation in the energy balance,

$$0 = \frac{1}{2g_c}G^2\left(\frac{RT}{M}\right)^2 d\left(\frac{z}{P}\right)^2 + \frac{RT}{M}\frac{z}{P}\,dP + \left[\frac{g}{g_c} + f\frac{G^2(RT/M)^2(z/P)^2}{2g_cD}\right]dL$$

or

$$0 = \left[\frac{1}{g_c}G^2\left(\frac{RT}{M}\right)^2\frac{z}{P}\left(\frac{1}{P}\frac{dz}{dP} - \frac{z}{P^2}\right) + \frac{RT}{M}\frac{z}{P}\right]dP$$

$$+ \left[\frac{g}{g_c} + f\frac{G^2(RT/M)^2(z/P)^2}{2g_cD}\right]dL.$$

Since the compressibility factor z in isothermal flow is a function only of P, and since f is a function of the Reynolds number $\text{Re} = DG/\mu$, where the viscosity μ becomes a function only of pressure or else a constant, then the preceding equation establishes a relationship between the pressure P and the distance L. Integration, however, will require numerical methods.

If the kinetic energy term is ignored,

$$0 = \frac{Rt}{M}\frac{z}{P}\,dP + \left[\frac{g}{g_c} + f\frac{G^2(RT/M)^2(z/P)^2}{2g_cD}\right]dL.$$

For an incompressible liquid, the relationship reduces to

$$0 = \frac{1}{\rho} dP + \left[\frac{g}{g_c} + f \frac{(G/\rho)^2}{2g_c D} \right] dL,$$

where ρ is constant. Changes in velocity do not occur for an incompressible fluid flowing through a constant diameter D.

In downflow, $-dz = dL$. This will change the sign of the term g/g_c in the preceding relationships.

Static Fluid Column. The preceding formulas are applicable to static conditions if $v = 0$. Thus for shut-in, only a vertical column of static fluid needs to be considered. Here, if the point elevation of fluid in the column is designated as z_{elev},

$$0 = \frac{g}{g_c} dz_{elev} + \frac{1}{\rho} dP.$$

From another viewpoint, this is the definition for pressure.

For a compressible fluid,

$$0 = \frac{g}{g_c} dz_{elev} + \frac{zRT}{MP} dP.$$

In general, say with a static petroleum or natural gas fluid such as exists in a well, temperature is a function of L or elevation—the temperature gradient. In turn, the compressibility factor z is a function of temperature and pressure. Thus numerical procedures will be required for the more general solution.

The acceleration of gravity is also a function of elevation, but variations may be neglected except under the more unusual circumstances—say, very high columns or very deep holes such as occur in deep oil and gas wells.

If the compressibility factor z and the temperature T are constant, then integrating between limits 1 and 2 gives the usual expression

$$\frac{P_2}{P_1} = \exp\left(- \frac{g}{g_c} \frac{M}{zRT} [(z_{elev})_2 - (z_{elev})_1] \right).$$

For an incompressible fluid, it is readily obtained that

$$P_2 - P_1 = - \frac{g}{g_c} \rho[(z_{elev})_2 - (z_{elev})_1],$$

where the density ρ is constant.

These formulas can be used to estimate the bottom-hole shut-in pressure of oil and gas wells under the appropriate circumstances.

Buoyancy. The movement of a solid body up or down through a fluid (a liquid or gas), or the flow of a fluid up or down around a suspended solid body, may also be handled by the energy balance relationship, whereby

$$d\bar{q} - d\bar{w} = \frac{1}{2g_c} dv^2 + \frac{g}{g_c} dz_{elev} + \bar{V}_B \, dP + d\overline{lw},$$

where the specific volume $\bar{V}_B$ of the body is the reciprocal of its density ρ_B; that is, the volume of the fluid system is equal to the volume of the body, per unit mass of the body, and the foregoing energy balance equation basis would be per unit mass of the body.

Moreover, it will be assumed first that the system is isolated with no exchange with the surroundings, and that no lost work or dissipative effects are involved. Ordinarily, it also can be assumed that $g = g_c$, although this simplification will not be made here.

Furthermore, for the vertical upward or downward flow of the fluid per se, where $dz = dz_{elev}$ and assuming no changes in the fluid velocity,

$$dP = -\rho_{fluid} \frac{g}{g_c} dz.$$

Combining the relationships, it follows that

$$-\frac{\rho_B}{2} \frac{dv^2}{dz} = g(\rho_B - \rho_{fluid})$$

or

$$-\frac{\rho_B}{2} \frac{dv^2}{dz} \frac{dt}{dt} = g(\rho_B - \rho_{fluid})$$

or

$$\rho_B \frac{dv}{dt} = g(\rho_B - \rho_{fluid}),$$

where $v = -dz/dt$.

For the force downward on the body, per unit mass of the body,

$$F_{down} = \frac{dv}{dt} = g\frac{\rho_B - \rho_{fluid}}{\rho_B},$$

and for the force upward per unit mass of the body,

$$F_{up} = -\frac{dv}{dt} = -g\frac{\rho_B - \rho_{fluid}}{\rho_B}.$$

If, however, dissipative or frictional effects are involved, they may be introduced into a force balance. It is the custom to introduce frictional or dissipative effects by the use of a friction factor f_D derivable via Stokes' law for flow around spheres; that is, in Stokesian flow, the dissipative force or resisting force $F = 3\pi D\mu v$, where $D = D_p$ is the particle diameter and μ is the fluid viscosity. Thus for a spherical body of diameter D or D_p,

$$\frac{\pi D^3}{6}\rho_B\frac{dv}{dt} = \frac{\pi D^3}{6}\rho_B g\frac{\rho_B - \rho_{fluid}}{\rho_B} - 3\pi D\mu v.$$

At the terminal velocity, $dv/dz = 0$. Therefore,

$$v = \frac{gD^2}{18\mu}(\rho_B - \rho_{fluid})$$

or

$$v^2 = \frac{(4/3)gD}{24/(Dv\rho_{fluid}/\mu)}\frac{\rho_B - \rho_{fluid}}{\rho_{fluid}},$$

where the convention used enables the substitution to be made such that the friction factor f_D is

$$f_D = 24/\frac{Dv\rho}{\mu}.$$

It follows that

$$v = \left[\frac{4(\rho_B - \rho_{fluid})gD}{3\rho_{fluid}(f_D)}\right]^{1/2}$$

and

$$f_D = \frac{2(\rho_B - \rho_{fluid})2g_c D}{3v^2\rho_{fluid}}.$$

For equilibrium, it would be required that the densities of the body and the fluid be equal.

Draft. For the draft in a vertical circular duct, chimney, or stack, consider the friction factor form

$$d\overline{w} = f\frac{v^2}{2g_c D}\,dL = f\frac{(G/\rho_{st})^2}{2g_c D}\,dL$$

such that at constant mass velocity, with no exchanges with the surroundings, the flow equation is

$$0 = \frac{g}{g_c}\,dz + \frac{1}{\rho_{st}}\,dP_{st} + f\frac{(G/\rho_{st})^2}{2g_c D}\,dL,$$

where $dL = dz = dz_{elev}$ and the subscript st stands for the stack gases, flue gases, or combustion products, such as they are or may be. Furthermore,

$$\rho_{st} = \frac{M_{st} P_{st}}{z_{st} RT_{st}} = \frac{P_{st}}{(zRT/M)_{st}}.$$

Integrating at a constant or mean density,

$$-\left[\frac{g}{g_c} + f\frac{(G/\rho_{st})^2}{2g_c D}\right]\rho_{st} = \frac{(P_{st})_2 - (P_{st})_1}{L_2 - L_1}.$$

For the ambient air A outside the stack, where no frictional effects are involved,

$$0 = \frac{g}{g_c}\,dz + \frac{1}{\rho_A}\,dP_A,$$

where

$$\rho_A = \frac{M_A P_A}{z_A RT_A} = \frac{P_A}{(zRT/M)_A}.$$

Integrating between limits at constant density, where $dz = dL$,

$$-\frac{g}{g_c}\rho_A = \frac{(P_A)_2 - (P_A)_1}{L_2 - L_1}.$$

Therefore, since the overall pressure drop for the stack gases and the air will be equal,

$$-\left[\frac{g}{g_c} + f\frac{(G/\rho_{st})^2}{2g_c D}\right]\rho_{st} = -\frac{g}{g_c}\rho_A.$$

Rearranging,

$$f\left(\frac{G}{\rho_{st}}\right)^2 = 2gD\left[-1 + \frac{\rho_A}{\rho_{st}}\right].$$

This result will permit a first estimation of the stack gas mass velocity G in terms of the density of the ambient air and the density of the stack gases, assuming average or mean temperatures and pressures for each. In general the friction factor f is a function of velocity (the Reynolds number); thus, it requires a trial-and-error solution. If f is assumed to be a constant or mean value, then the velocity calculates directly as a square root. Alternately, Poiseuillean flow may be assumed, whereby $f = 64/\text{Re} = 64/(DG/\mu)$ such that

$$\frac{64G\mu}{D\rho_{st}^2} = 2gD\left[-1 + \frac{\rho_A}{\rho_{st}}\right].$$

This will permit solving directly for G.

More exactly, the integrations should be at varying gas densities, at least with respect to pressure, albeit a mean temperature may be assumed. The determination will be much more complicated and will include the stack height, as follows.

For the stack gases, substituting for the density,

$$-\int_{(P_{st})_1}^{(P_{st})_2}\left\{[(zRT/M)_{st}1/P_{st}]\middle/\left(\frac{g}{g_c} + \frac{fG^2(zRT/M)_{st}1/P_{st}^2}{2g_cD}\right)\right\} dP_{st} = L_2 - L_1$$

is obtained. If f is constant or if Poiseuillean flow is assumed, by multiplying the numerator and denominator through by P_{st}^2, the preceding integral will have an analytic solution of the type

$$\int \frac{x}{ax^2 + c} dx = \frac{1}{2a}\ln(ax^2 + c).$$

For the air,

$$-\int_{(P_A)_1}^{(P_A)_2} \frac{(zRT/M)_A 1/P_A}{g/g_c} dP_A = L_2 - L_1$$

and

$$\frac{(P_A)_2}{(P_A)_1} = \exp\left[-\frac{g/g_c}{(zRT/M)_A}(L_2 - L_1)\right]$$

or

$$(P_A)_2 - (P_A)_1 = (P_A)_1\left\{\exp\left[-\frac{g/g_c}{(zRT/M)_A}(L_2 - L_1)\right] - 1\right\}.$$

For a specified stack height $L_2 - L_1$ and given, say, $(P_A)_1$, then $(P_A)_2$ can be determined from the foregoing equations for the air. In addition, since by definition $(P_{st})_1 = (P_A)_1$ and $(P_{st})_2 = (P_A)_2$, the mass velocity G can be determined from the integral for the stack gases—albeit this will be trial-and-error. The particular solution obtained will depend upon the behavior of the friction factor f.

In the preceding relationships, note that the kinetic energy changes are not included. For incompressible liquids, the velocity is constant at steady state and there would be no kinetic energy change. For gases, however, first there is the consideration that including kinetic energy changes would vastly complicate the solutions. More significantly, there is the fact that for gases, the kinetic energy changes would be small compared to the pressure–volume energy change. Thus consider

$$\frac{dv^2}{2g_c} = \frac{d(G/\rho)^2}{2g_c} = \frac{G^2(zRT/M)^2 d(1/P)^2}{2g_c} = \frac{G^2(zRT/M)^2}{g_c(-P^3)}dP.$$

For most purposes,

$$\left|\frac{G^2(zRT/M)^2}{g_c(-P^3)}\right|dP < \bar{V}\,dP = \frac{zRT/M}{P}\,dP$$

or

$$\frac{v^2}{g_c} < \frac{zRT}{M}\frac{P}{P} = \frac{P}{\rho},$$

where the units are to be consistent.

Siphon. For a siphon, there is no pressure change. Accordingly,

$$0 = \frac{g}{g_c} \, dz_{elev} + f \frac{(G/\rho)^2}{2g_c D} \, dL.$$

The elevation change will be less than the distance change. More applicable to liquids, the mass velocity G is constant at steady state.

Thermosiphon. In the upflow of a fluid through a heated tube, at a constant mass velocity, it may be assumed that

$$dq = \frac{dq}{dL} \, dL = \frac{g}{g_c} \, dz_{elev} + \frac{1}{\rho} \, dP + \frac{f(G/\rho)^2}{2g_c D} \, dL,$$

where $dZ_{elev} = dL$ and dq/dL is het added per linear distance. This incorporates an additional term into the energy balance, whereby

$$-\frac{1/\rho}{g/g_c - dq/dL + \left[f(G/\rho)^2 \right]/2g_c D} \, dP = dL.$$

This relationship may be integrated because the behavior of dq/dL is known and may be assumed a constant or mean value. For the purposes of integration the temperature of the fluid may be assumed constant. For a liquid, the density may be assumed constant, whereas for a gas it will be a function of pressure, at a constant or mean temperature.

Pipeline Equation. In horizontal flow, $dz_{elev} = 0$ and the potential energy term would be cropped from the energy balance. In general, however, z_{elev} may be made a function of the distance L, which leads to the pipeline equation. For isothermal flow with no work transfer,

$$0 = \frac{d(G/\rho)^2}{2g_c} + \frac{g}{g_c} \frac{dz_{elev}}{dL} \, dL + \frac{1}{\rho} \, dP + f \frac{(G/\rho)^2}{2g_c D} \, dL.$$

For a compressible gas,

$$0 = \left[\frac{1}{g_c} G^2 \left(\frac{RT}{M} \right)^2 \frac{z}{P} \left(\frac{1}{P} \frac{dz}{dP} - \frac{z}{P^2} \right) + \frac{RT}{M} \frac{z}{P} \right] dP$$

$$+ \left[\frac{g}{g_c} \frac{dz_{elev}}{dL} + f \frac{G^2 (RT/M)^2 (z/P)^2}{2g_c D} \right] dL.$$

Neglecting the kinetic energy change,

$$0 = \frac{RT}{M} \frac{z}{P} dP + \left[\frac{g}{g_c} \frac{dz_{\text{elev}}}{dL} + f \frac{G^2(RT/M)^2(z/P)^2}{2g_c D} \right] dL.$$

For an incompressible fluid,

$$0 = \frac{1}{\rho} dP + \left[\frac{g}{g_c} \frac{dz_{\text{elev}}}{dL} + f \frac{(G/\rho)^2}{2g_c D} \right] dL.$$

The behavior of the derivative dz_{elev}/dL is required before integration or solution can proceed.

1.6. FLOW RATE

The energy balance can be adapted to the measurement of flow rate for both liquids and gases by relating velocity and pressure–volume effects. A discharge coefficient is introduced, by which the one can be correlated to the other. (Positive displacement and mass flow meters are excluded.)

Orifices, Venturis, and Pitots

The orifice equation incorporates the effect of the lost work term lw into a coefficient. Thus, for horizontal flow, it is assumed that

$$0 = \frac{1}{2g_c} dv^2 + C_o^2 V dP,$$

where C_o^2 is the coefficient introduced. Integrating,

$$v_2^2 - v_1^2 = C_o^2 \cdot 2g_c(-)\int_{P_1}^{P_2} V \, dP.$$

Since $v = W/\rho A$, where W is the mass velocity and A is the cross-sectional area, then for a contraction from A_1 to A_2 in the direction of flow,

$$W = C_o \sqrt{\frac{2g_c(-)\int_1^2 V \, dP}{(1/A_2)^2 - (1/A_1)^2}}.$$

This, in all its variations, constitutes the orifice equation. Correlations for the discharge coefficient C_o are presented in Fig. 1.3. These correlations

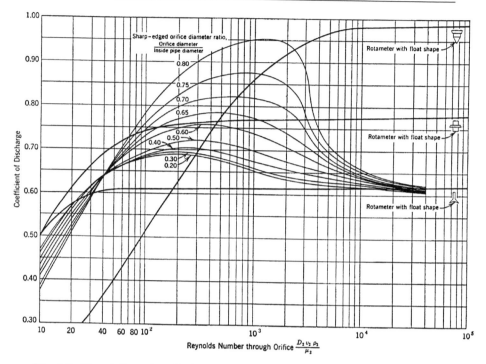

Fig. 1.3. Correlations for discharge coefficients. (Courtesy of the Fischer and Porter Company, Warminster PA. Reproduced in G. G. Brown *et al. Unit Operations*, Wiley, New York, 1950.)

apply to orifice designs and orifice runs as specified by the American Society of Mechanical Engineers (5, 6).

For well-designed venturis, the coefficient C_o is close to unity (~ 0.98).

For pitot tubes, the velocity head is changed to a pressure–volume head. Thus

$$ v = C_o \sqrt{2g_c(-)\int_{P_1}^{P_2} V\,dP}\,. $$

For the small changes which occur,

$$ v = C_o \sqrt{2g_c \frac{P_1 - P_2}{\rho}}\,. $$

The coefficient C_o is usually accepted as unity.

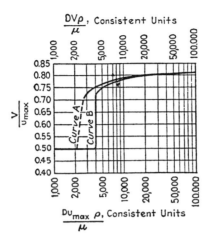

Fig. 1.4. Pipe factor in circular pipes. (From *Heat Transmission*, 2nd ed., by W. H. McAdams. Copyright 1942 by McGraw-Hill. Used with permission of McGraw-Hill Book Company.)

If v is in feet per second, where $g_c = 32.2$ ft/sec, and the pressure drop is measured in pounds per square foot (psf), then

$$P_1 - P_2 = \Delta h \frac{62.4}{12},$$

where Δh is in inches of water and the density of water is set at 62.4 lb/ft^3.

Pitot readings are usually taken in the center of the pipe. This is a maximum reading and should be adjusted to obtain the average reading as shown in Fig. 1.4 in terms of what is called the pipe factor (7). Alternatively, a traverse may be run.

1.7. ISENTROPIC BEHAVIOR AND EFFICIENCY

The generalized energy balance applies for compression and expansion as well as for fluid flow. Thus, for instance, in consistent mechanical units, per unit mass of fluid,

$$dq - dw = \frac{1}{2g_c} dv^2 + \frac{g}{g_c} dz + V\,dP + d\omega,$$

where, if $d\omega = d\text{lw} = T dS = 0$, the behavior is called isentropic. If the behavior also is adiabatic, then $dq = 0$. (It is "isentropic" as distinguished from "isotropic"; the latter pertains to a physical property that has the same value along each axis.)

The preceding expression is on a unit mass basis or on a unit mass flow rate basis. Bars above the variables are customarily used to indicate this fact, as previously noted, but may be dropped for convenience provided that such information is understood. If velocity and elevation effects are not to be included, then the expression may be placed on a molar or other basis. In any case, the units must be consistent, by definition.

Neglecting velocity and elevation changes, and assuming no heat transfer, for an open or flow system in mechanical units,

$$-dw = V dP + d\omega = J_0 dH,$$

where H is in heat units. As previously indicated, the mechanical equivalent of heat J_0 may be dropped if it is understood that the units are to be consistent.

The expression may be integrated if ω is known as a function of P. The usual procedure is to use an efficiency correction instead. Therefore, let

$$-dw_{\text{theo}} = V dP$$

and

$$-dw = \frac{-dw_{\text{theo}}}{\text{Eff}} \quad \text{or} \quad -dw_{\text{theo}} = \text{Eff}(-dw).$$

Since

$$V dP + d\omega = J_0 dH = J_0[C_p dT - C_p \mu dP],$$

this is an additional relationship based on $d\omega$ and μ. If μ is assigned the value zero and $d\omega$ is neglected, then for an ideal gas undergoing an isentropic change,

$$V dP = J_0 C_p dT$$

or

$$\frac{RT}{P} dP = J_0 C_p dT.$$

Integration between limits will yield the customary form,

$$\frac{T_2}{T_1} = \left(\frac{P_2}{P_1}\right)^{R/J_0 C_p} = \left(\frac{P_2}{P_1}\right)^{R_0/C_p} = \alpha,$$

where α is defined by the preceding relationship. Furthermore,

$$-dw = J_0 C_p \, dT \quad \text{(in mechanical units)}$$

or, if we so choose,

$$-dW = C_p \, dT \quad \text{(in heat units)}.$$

Most often, work will be expressed in lowercase characters and R will symbolize the gas constant in both mechanical and heat units, with the understanding that the units are to be consistent.

Alternatively, for a closed system, it will follow that

$$-P \, dV = J_0 C_v \, dT,$$

from which

$$\frac{-RT}{V} \, dV = J_0 C_v \, dT$$

and

$$\frac{T_2}{T_1} = \left(\frac{V_2}{V_1}\right)^{-R/J_0 C_v} = \left(\frac{V_2}{V_1}\right)^{-R_0/C_v} = \alpha,$$

where R_0 is in heat units.

Finally, since

$$V \, dP = C_p \, dT \quad \text{and} \quad -P \, dV = C_v \, dT,$$

then

$$d(PV) = V \, dP + P \, dV = (C_p - C_v) \, dT = R \, dT.$$

Therefore, the usual result is obtained:

$$C_p - C_v = R \text{ or } R_0,$$

where the units are to be consistent. It will follow also that

$$\frac{T_2}{T_1} = \left(\frac{P_2}{P_1}\right)^{(C_p - C_v)/C_p} = \left(\frac{P_2}{P_1}\right)^{(k-1)/k} = \alpha,$$

where $k = C_p/C_v$.

Alternatively, in terms of V and P, on substituting the perfect gas law for T,

$$\frac{V_2}{V_1} = \left(\frac{P_2}{P_1}\right)^{-1/k}$$

or

$$P_2 V_2^k = P_1 V_1^k,$$

etc.

These ideal compressions or expansions are called isentropic since it is assumed that $d\omega = d\text{lw} = T\,dS = 0$. Efficiency ratings are specified on the basis of the calculated isentropic energy change.

Isentropic Efficiency. For an isentropic compression, it may be written over the interval that, in consistent units,

$$-w_{\text{act}} = \frac{-w_{\text{theo}}}{\text{Eff}} = \frac{C_p(T_2 - T_1)}{\text{Eff}} = \frac{C_p T_1 (T_2/T_1 - 1)}{\text{Eff}},$$

where the ratio T_2/T_1 is determined for isentropic behavior and where Eff denotes an overall fractional efficiency. Both w_{theo} and w_{act} are negative, which denotes work done on the system.

For an isentropic expansion, on the other hand,

$$-w_{\text{act}} = \text{Eff}(-w_{\text{theo}}) = \text{Eff} \cdot C_p(T_2 - T_1) = \text{Eff} \cdot C_p T_1 \left[\frac{T_2}{T_1} - 1\right],$$

where $T_1 > T_2$ and the ratio T_2/T_1 is again determined from isentropic behavior. It will be observed that the convention is such that w_{theo} and w_{act} are both positive for work done by the expansion. (To be more consistent, capital "W" could be used to denote heat units.)

Efficiency and Nonisentropic Behavior

In general we write, in consistent units,

$$dq - dw = dH = V\,dP + d\omega,$$

where $d\omega = J_0\,d\Omega = d\text{lw} = T\,dS$ is the intrinsic energy change or lost work change (or dissipative energy change or the irreversibility). These relationships may be adapted so as to introduce a pointwise efficiency ϵ for either compression or expansion.

Compression. Parts of the foregoing expression may be arranged to read, in consistent units,

$$\epsilon \, dH = V \, dP,$$

where also

$$(1 - \epsilon) \, dH = d\omega \quad \text{or} \quad \frac{1 - \epsilon}{\epsilon} V \, dP = d\omega.$$

Here the symbol ϵ denotes the pointwise efficiency for a nonisotropic compression and the terms are all positive.

Assuming that $dH \sim C_p \, dT$, it will follow that

$$\frac{T_2}{T_1} = \left(\frac{P_2}{P_1} \right)^{(k-1)/\epsilon_m k},$$

where $T_2 > T_1$ and where ϵ_m represents a mean or average value over the interval of integration. Furthermore, if $q = 0$, then for compression,

$$-w_{\text{act}} = C_p(T_2 - T_1) = C_p T_1 \left[\frac{T_2}{T_1} - 1 \right],$$

where T_2/T_1 represents the actual absolute temperature ratio as determined previously. (We note that $-w_{\text{act}}$ is used for the integrated value, although we could as well use Δw.)

Note that in compression, for a given compression pressure ratio, the actual temperature or temperature ratio achieved will be greater than the theoretical value for isentropic compression, due to the effect of the irreversibilities or efficiency losses that show up as waste heat. In turn, the work requirement is greater than the theoretical value calculated for isentropic compression.

Also note that, over the interval and in consistent units,

$$(1 - \epsilon_m) \Delta H = \Delta \omega,$$

which provides a value for the extent of the irreversibility or lost work.

Expansion. In an expansion, where again

$$dq - dw = dH = V \, dP + d\omega$$

and where both dH and $V dP$ are negative, a pointwise efficiency ϵ may be introduced as

$$dH = \epsilon V dP,$$

where here

$$-(1 - \epsilon)V dP = d\omega \quad \text{or} \quad \frac{-(1 - \epsilon)}{\epsilon} dH = d\omega.$$

Then, for an expansion at a mean efficiency ϵ_m,

$$\frac{T_2}{T_1} = \left(\frac{P_2}{P_1}\right)^{\epsilon_m(k-1)/k},$$

where the convention used is that $T_1 > T_2$ in the direction of flow. If $q = 0$, then again

$$-w_{\text{act}} = C_p(T_2 - T_1) = C_p T_1\left[\frac{T_2}{T_1} - 1\right].$$

It is observed that the ratio T_2/T_1 is larger than would be calculated for an isentropic expansion for the same pressure expansion ratio; that is, given T_1, the downstream expansion temperature T_2 would be larger than that calculated for a theoretical isentropic expansion. Moreover, the value of the work delivered, w_{act}, would be smaller than the theoretical value, which indicates that irreversibilities occur. Furthermore,

$$\frac{-(1 - \epsilon_m)}{\epsilon_m} \Delta H = \Delta \omega,$$

which provides an indication of the degree of irreversibility. Here, of course, ΔH is negative.

1.8. NONADIABATIC BEHAVIOR

In consistent (say mechanical) units, and neglecting kinetic and potential energy effects, the energy balance per unit mass for a flow system arranges to

$$-dw = dH - dq = V dP + (d\omega - dq) = V dP + d\omega^*,$$

where

$$d\omega^* = d\omega - dq \quad (\text{or in heat units, } d\Omega^* = d\Omega - dQ)$$

will here be called the generalized irreversibility. For isothermal flow it would be equivalent to $d\text{lw}_{\text{isothermal}}$. Of principal interest here, however, is nonadiabatic compression or expansion with work exchange with the surroundings.

Observe that if isentropic behavior is assumed, where $d\omega = d\text{lw} = TdS = 0$, then $dH = VdP$, which reduces to $C_p dT = VdP$ for a perfect gas, and the usual form for an isentropic compression or expansion will again be attained, since dq will cancel out. Thus there is no provision for nonadiabatic isentropic behavior in adjusting the temperature-pressure locus, nor for adiabatic nonisentropic behavior. The only provision is for adiabatic isentropic behavior, and thus the latter descriptors in a sense become redundant. Isentropic behavior therefore implies adiabatic behavior and vice versa.

We also may adopt the notation that, in consistent units,

$$d\omega^* \text{ or } d\Omega^* = d\text{lw}^* = T\,dS^*.$$

Here, for generalized "isentropic" behavior,

$$d\omega^* \text{ or } d\Omega^* = 0, \qquad d\text{lw}^* = 0, \qquad dS^* = 0.$$

Therefore, if it is assumed that, for the generalized irreversibility,

$$d\omega^* = 0 \quad \text{or} \quad d\omega = dq,$$

then the effect of nonadiabatic behavior can be introduced. This qualification can be referred to as reversible nonadiabatic behavior. (A special case of the preceding condition is if both $d\omega = 0$ and $dq = 0$, then $d\omega^* = 0$. It then becomes the situation of isentropic adiabatic change.)

There will be obtained

$$-dw = dH - dq = V\,dP,$$

whereby, for a perfect gas,

$$C_p\,dT - dq = \frac{RT}{P}\,dP.$$

If, say, q is made some arbitrary function of temperature, $q = q(T)$, then let the derivative be represented symbolically as

$$\frac{dq}{dT} = f = f(T)$$

and it follows that

$$\frac{C_p - f}{T} dT = \frac{R}{P} dP.$$

The integration or solution would yield T versus P. In turn,

$$-dw = (C_p - f)\, dT = \frac{RT}{P} dP.$$

Integration will yield $-w$ in terms of T or P. In this way provision can be made to introduce the effects of nonadiabaticity.

Note furthermore that for, say, heat *removal* during compression, the function f will be negative, and since

$$dT = \frac{RT}{C_p - f} \frac{dP}{P},$$

this will, as expected, cause the temperature to rise more slowly than during an isentropic compression, where in the latter case,

$$dT = \frac{RT}{C_p} \frac{dP}{P}.$$

Moreover, the pressure–volume work term $(-dw)$, the work input required for compression, will be smaller; that is,

$$-dw = \frac{RT}{P} dP$$

will be smaller for a given pressure change, since the temperature rises more slowly than for isentropic compression; that is, $(-w)$ will be less.

On the other hand, for a given temperature change, using the form

$$-dw = (C_p - f)\, dT,$$

the work term $(-w)$ will be larger, since $(C_p - f)\, dT$ will rise faster than $C_p\, dT$. Thus the compression requirement would be larger in this latter instance.

The effect of *adding* heat may be investigated similarly. Work exchange during nonadiabatic expansion—that is, through a mechanical expander or turbine—also may be treated in the same fashion, where now both dT and dP are regarded as negative in the direction of flow. The function f,

$$\frac{dq}{dT} = f = f(T),$$

will again be negative, since in this case dq is positive and dT is negative.

It will be found that for a given expansion pressure ratio, the temperature will fall more slowly than for an adiabatic isentropic expansion, but the work done will be greater than for an adiabatic isentropic expansion; that is, the temperature remains correspondingly higher than during an adiabatic isentropic expansion, such that the integral of

$$dw = \frac{-RT}{P}\,dP$$

will be greater.

Alternatively, heat losses (or additions) may be adjusted into the isentropic solutions by the use of an efficiency factor in the manner described and developed in the previous section.

Interstage cooling (or heating) between successive isentropic adiabatic compressions (or expansions) is another means to accommodate nonadiabatic behavior and will be deployed extensively in Chapter 4. Interstage cooling is the more practical embodiment because it is allied with actual equipment.

1.9. NOZZLES

For a reversible change in an ideal gas,

$$V\,dP = C_p\,dT = \frac{C_p}{R}\,d(PV) = \frac{C_p}{C_p - C_v}\,d(PV) = \frac{k}{k-1}\,d(PV).$$

For an isolated flow system of unit mass, the energy balance will reduce to

$$0 = \frac{1}{2g_c}\,dv^2 + V\,dP = \frac{1}{2g_c}\,dv^2 + \frac{k}{k-1}\,d(PV).$$

This will integrate and rearrange to

$$\frac{1}{2g_c}v^2 = \frac{k}{k-1}[P_0V_0 - PV],$$

where the nozzle inlet conditions are

$$P = P_0 \quad \text{and} \quad V = V_0 \quad \text{at } v = 0.$$

This is the static condition for the inlet.

If a critical velocity v_c occurs at P_c, V_c, T_c and the expansion through the nozzle obeys the relationships for isentropic compression or expansion, then

$$P_cV_c^k = P_0V_0^k$$

and

$$v_c^2 = kg_c P_0 V_0 \left\{ \frac{2}{k-1} \left[1 - \left(\frac{P}{P_c} \right)^{(k-1)/k} \right] \right\}.$$

For air, the critical pressure ratio P_c/P_0 is about 0.52 and $k \sim 1.403$. It follows that

$$\frac{2}{k-1}[1 - (0.52)^{(k-1)/k}] \sim 1.0,$$

from which

$$v_c^2 = kg_c P_0 V_0.$$

It will follow also that v_c is approximately the speed of sound in air if consistent units are used:

$$v_c = \sqrt{1.403(32.2)P\left(\frac{RT_0}{MP_0}\right)},$$

where the molecular weight of air is $M = 29$. At standard conditions, where $P = P_0$,

$$v_c = \sqrt{1.403(32.2)(1543)\frac{460 + 32}{29}} = 1087.5 \text{ ft/sec.}$$

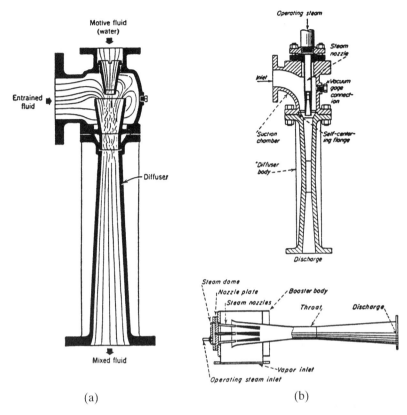

Fig. 1.5. Representative schematics for ejector–injectors. (a) Water-jet ejector (Schutte and Koerting Co.). (b) Steam-jet ejector (*Chemical Engineers' Handbook*, 3rd ed., J. H. Perry, ed., McGraw-Hill, New York, 1950, p. 1453).

The velocity of sound in dry air at 1 atm and 0°C or 32°F is 1087.1 ft/sec. Thus the velocity at the throat is approximately the speed of sound. As the nozzle diverges, velocities much greater than the speed of sound are obtained, with attendant losses in pressure–volume effects. This phenomenon also is related to the Joule–Thomson coefficient.

Diagrams for nozzles are shown in Fig. 1.5 as applied to ejector–injectors (3, 8). The characteristics of a reversible adiabatic expansion are diagrammed in Figs. 1.6 and 1.7 (9, 10).

The accelerated flow through the nozzle produces a lowering of the pressure, which can be used to pick up a secondary stream. When two streams are involved, the total nozzle assembly is called an ejector or an

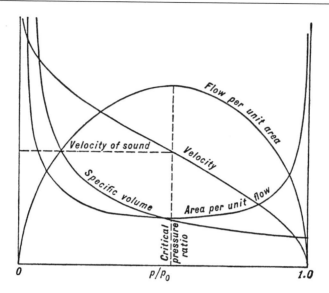

Fig. 1.6. Characteristics of a reversible adiabatic expansion. (From J. H. Keenan, *Thermodynamics*, Wiley, New York, 1941, p. 137.)

injector, depending upon which stream is of interest. The terms adductor and inductor may be used also.

Nozzles do not have to operate at sonic or supersonic conditions to have an effect on pressure. Furthermore, compressible fluids (liquids) may be employed for either or both of the streams.

Ejector–Injector Analysis

An analysis of the combining of two streams involves a total energy balance. The sequence is diagrammed as follows:

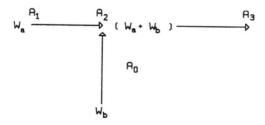

Primary stream a at a constant mass flow rate W_a is constricted from a cross-sectional area A_1 and velocity v_1 at pressure P_1 down to area A_2 and velocity v_2 at pressure P_2. Secondary stream b at a constant mass flow

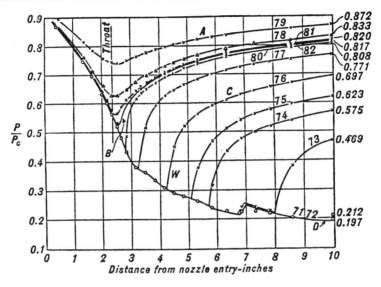

Fig. 1.7. Pressure distribution in a converging–diverging nozzle expressed in terms of the ratio of pressure to upstream or entrance pressure for various downstream or exit pressures. [From A. M. Binney and M. W. Woods, *Proc. Inst. Mech. Eng.* *138*, 260 (1938). Reproduced in J. H. Keenan, *Thermodynamics*, Wiley, New York, 1941.]

rate W_b is constricted from a cross-sectional area A_0 and velocity v_0 down to area A_2 and velocity v_2 at pressure P_2. At point 2 or area A_2 the streams mix. The mixed streams a plus b at A_2 are expanded downstream to area A_3 at pressure P_3. A total energy balance at the point of mixing is

$$0 = W_a \left\{ \frac{v_2^2 - v_1^2}{2g_c} + \int_{P_1}^{P_2} \frac{dP}{\rho_a} + \text{lw}_a \right\}$$
$$+ W_b \left\{ \frac{v_2^2 - v_0^2}{2g_c} + \int_{P_0}^{P_2} \frac{dP}{\rho_b} + \text{lw}_b \right\},$$

where lw represents dissipative energy changes that occur in the direction of flow for the designated stream. If $W_b = 0$, the relationship becomes that of the single stream a. Coefficients C_a and C_b may be introduced ahead of the integrals (ideally equal to unity).

For the expansion of the combined streams,

$$0 = (W_a + W_b) \left\{ \frac{v_3^2 - v_2^2}{2g_c} + \int_{P_2}^{P_3} \frac{dP}{\rho_{a+b}} + \text{lw}_{a+b} \right\}$$

or

$$0 = \frac{v_3^2 - v_2^2}{2g_c} + C_{a+b} \int_{P_2}^{P_3} \frac{dP}{\rho_{a+b}}.$$

For most practical purposes, lw $= 0$ and $C = 1$.

All but two quantities and/or limiting values may be assigned. The other two are, in principle, determinable from the preceding two independent equations. The problem may be stated several ways: some statements are more convenient than others. Since two equations are involved, the choice of the two variables to be solved for is arbitrary, but practically speaking there is the matter of ease of solution. Thus, say, given the dimensions, the primary flow rate, and the inlet and outlet pressures (P_1 and P_3), it is possible to solve for the secondary flow rate (and the pressure at the nozzle throat P_2). An example is provided in reference 11.

If a and b are immiscible phases, then two-phase flow will result, but the principles remain the same. The coefficients would be expected to change, however.

1.10. DISCHARGE COEFFICIENTS VERSUS DISSIPATIVE EFFECTS

The discharge coefficient may be related directly to lost work or dissipative effects, and it becomes a measure of efficiency.

Efficiency

The discharge coefficient C_o for an orifice, venturi, pitot, or nozzle enters the energy balance, for isothermal flow, is

$$0 = \frac{1}{2g_c} dv^2 + C_o^2 V dP.$$

Alternately, since the energy balance may also be represented as

$$0 = \frac{1}{2g_c} dv^2 + V dP + d\text{lw},$$

it follows that

$$0 = (1 - C_o^2) V dP + d\text{lw}$$

or

$$dlw = (1 - C_o^2)(-VdP).$$

On integrating, for liquids,

$$lw = \Delta lw = (1 - C_o^2)\frac{1}{\rho}(P_1 - P_2).$$

For gases,

$$lw = \Delta lw = (1 - C_o^2)\frac{zRT}{MW}\ln\frac{P_1}{P_2},$$

where z is the mean or average compressibility factor and MW or M is the molecular weight.

An efficiency rating may in turn be defined by

$$\text{Eff} = C_o^2 \quad \text{or} \quad \text{Eff loss} = (1 - C_o^2).$$

Values of C_o are about 0.61 for orifice vena contracta taps at high Reynolds numbers, and up to 0.98 or 0.99 for well-designed venturis. Pitot tubes approach unity. Nozzles may have values of 0.9 or higher, although at supersonic velocities the value will fall off due to additional irreversibilities which occur.

From another standpoint, for an isolated flow system,

$$0 = \frac{1}{2g_c}dv^2 + VdP + dlw_{\text{isolated}}$$

or

$$0 = \frac{1}{2g_c}dv^2 + B_o^2 VdP,$$

where B_o is the discharge coefficient for an isolated system and where it is required that

$$dlw = dlw_{\text{isothermal}} = dlw_{\text{isolated}} - dq.$$

It follows that

$$dlw_{\text{isolated}} = (1 - B_o^2)(-VdP).$$

For a perfect gas in an isentropic expansion, in consistent units,

$$V \, dP = C_p \, dT \quad \text{and} \quad \int V \, dP = C_p(T_2 - T_1).$$

Substituting the perfect gas law $PV = RT$ and integrating to the customary form,

$$\frac{T_2}{T_1} = \left(\frac{P_2}{P_1}\right)^{(k-1)/k}$$

or

$$T_2 - T_1 = T_1\left[\left(\frac{P_2}{P_1}\right)^{(k-1)/k} - 1\right].$$

Therefore,

$$\text{lw}_{\text{isolated}} = \Delta \text{lw}_{\text{isolated}} = (1 - B_o^2)C_p T_1\left[1 - \left(\frac{P_2}{P_1}\right)^{(k-1)/k}\right],$$

where the units are consistent and where B_o is assumed constant for the integration.

More precisely,

$$dH = V \, dP + d\text{lw}_{\text{isolated}} = C_p \, dT - C_p \mu \, dP,$$

where

$$d\text{lw}_{\text{isolated}} = d\omega$$

and

$$PV = \frac{z}{\text{MW}} RT.$$

Since

$$d\text{lw}_{\text{isolated}} = (1 - B_o^2)(-V \, dP),$$

then

$$-V \, dP + C_p \, dT - C_p \mu \, dP = (1 - B_o^2)(-V \, dP)$$

or

$$C_p \, dT = \left[C_p \mu + B_o^2 V \right] dP = \left[C_p \mu + B_o^2 \frac{z}{MW} \frac{RT}{P} \right] dP.$$

Solution of this differential equation establishes the relationship between T and P for parameters of B_o. Both μ and z are, in general, functions of T and P. Therefore,

$$d\mathrm{lw}_{\text{isolated}} = (1 - B_o^2) - \frac{z}{MW} \frac{RT}{P} \, dP.$$

Integration yields

$$\mathrm{lw}_{\text{isolated}} = \Delta \mathrm{lw}_{\text{isolated}} = -(1 - B_o^2) \frac{R}{MW} \int_1^2 \frac{zT}{P} \, dP,$$

where the limits of integration are from P_1 to P_2, and where the compressibility factor z is the function $z = z(T, P)$ and T is a function of P, that is, $T = T(P)$, by virtue of the solution of the previous differential equation. For convenience, z may be assumed a constant or mean value, and for a perfect gas, of course, $z = 1$.

Generalized Orifice Flow

By including kinetic energy effects, the most general form for horizontal steady-state flow through orifices or venturis, etc., may be represented by the overall energy balance

$$dq - dw = \frac{1}{2g_c} dv^2 + dH = \frac{1}{2g_c} + V \, dP + d\omega,$$

where units are consistent and where usually it is assumed that $d\omega = d\mathrm{lw} = 0$.

It will so happen that dv^2 initially will be positive and dP negative, whereas $d\omega$ will be positive. The heat exchange with the surroundings or heat added to the system (dq) may be positive or negative. For gases, however, the Joule–Thomson coefficient $\mu = dT/dP$ is generally positive at most conditions, indicating that dT will be negative, so that dq most likely will be positive. For liquids, the inverse holds, although $\mu = dT/dP$ will be only a very small negative value, such that the system can be regarded as essentially isothermal. If the system is adiabatic, then of course $dq = 0$. To continue, a rearrangement gives

$$0 = \frac{1}{2g_c} dv^2 + \left[1 + \frac{1}{V} \frac{d\omega}{dP} - \frac{1}{V} \frac{dq}{dP} \right] V dP,$$

where the term in brackets can be generalized to the square of a discharge coefficient of one type or another, depending upon the flow system. Thus for, say, isothermal flow, it can be written that

$$C_o^2 = \left[1 + \frac{1}{V} \frac{d\omega}{dP} - \frac{1}{V} \frac{dq}{dP} \right]_{\text{isothermal}},$$

where $d\omega/dP$ will be negative, and where dq/dP most likely also will be negative for gases and near zero or slightly positive for liquids.

The foregoing expression can be further modified to read

$$C_o^2 = \left[1 + \frac{1}{\overline{V}} \frac{d\omega_{\text{isothermal}}}{dP} \right],$$

where

$$d\omega_{\text{isothermal}} = d\omega - dq_{\text{isothermal}}.$$

For adiabatic flow, where $dq = 0$, and which in this case is also isolated flow, such that $d\omega = d\omega_{\text{isolated}} = d\omega_{\text{adiabatic}}$, it can be written that

$$B_o^2 = \left[1 + \frac{1}{V} \frac{d\omega_{\text{isolated}}}{dP} \right],$$

where $d\omega_{\text{isolated}}/dP$ is negative. Note that $d\omega = d\omega_{\text{isolated}}$ represents the true behavior of the intrinsic energy change

$$d\omega = d\text{lw} = T \, dS.$$

In principle, at least, the behavior of the discharge coefficient can be used to determine the behavior of the intrinsic energy, lost work, or entropy.

Finally, since

$$dq = \frac{1}{2g_c} dv^2 + dH \quad \text{or} \quad 0 = \frac{1}{2g_c} dv^2 + dH - dq,$$

then flow through orifices, etc., can be related to enthalpic and Joule–Thomson behavior, that is, for instance,

$$dH = \left[1 + \frac{1}{v}\frac{d\omega}{dP}\right]V\,dP = B_o^2 V\,dP$$

or

$$C_p\,dT - C_p\,\mu\,dP = B_o^2 V\,dP$$

or

$$C_p\,dT = \left[C_p\,\mu + B_o^2\right]V\,dP.$$

In terms of the derivative,

$$\frac{dT}{dP} = \frac{1}{C_p}\left[C_p\,\mu + B_o^2\right]$$

or

$$\mu = \frac{dT}{dP} - \frac{1}{C_p}B_o^2,$$

which relates the change in temperature with respect to pressure in the orifice, where velocity effects also are incurred. Note, therefore, that the preceding derivative dT/dP, is not the true Joule–Thomson coefficient, since velocity effects have been included by virtue of the isolated or adiabatic discharge coefficient B_o or B_o^2. The preceding expression, however, can serve as a correction between the experimental value dT/dP obtained and the true Joule–Thomson coefficient μ.

1.11. MULTIPHASE FLOW

The previously derived energy balances for the single-phase flow of gases or liquids may be adapted to two-phase concurrent flow, at both constant rates and varying rates (e.g., condensation and vaporization). The subject is, for the most part, outside of the intended purposes here, but a few observations are in order.

Boundary Layer Proration

The concept of boundary layer analysis provides a convenient means to introduce correlating forms for pressure-drop in two-phase flow. Each phase may be regarded as contributing a boundary layer with the appropriate viscous forces depending upon the flow rates and properties of the respective phases. The same reasoning may be extended to more than two phases, but for convenience can be confined to a heterogeneous system of only two phases.

The two-phase behavior is referenced to the behavior for single-phase flow, and the corresponding pressure-drops are prorated on a mass fraction basis. The subject is further developed in reference 4. The vertical and horizontal flow of two-phase gas–solid and liquid–solid mixtures is covered, for instance, by Govier and Aziz (12) who also cover aspects of gas–liquid and liquid–liquid systems in vertical and horizontal flow, including slugging. The phenomena become increasingly complex in all their ramifications.

REFERENCES

1. Hoffman, E. J., *Analytic Thermodynamics: Origins, Methods, Limits, and Validity*, Taylor & Francis, New York, 1991.
2. Moody, L. F., *Trans. ASME 68*, 672 (1944).
3. Brown, G. G., A. S. Foust, D. L. Katz, R. Schneidewind, R. R. White, W. P. Wood, G. M. Brown, L. E. Brownell, J. J. Martin, G. B. Williams, J. T. Banchero, and J. L. York, *Unit Operations*, Wiley, New York, 1950.
4. Hoffman, E. J., *Heat Transfer Rate Analysis*, PennWell, Tulsa, OK, 1980.
5. American Society of Mechanical Engineers, Special Committee on Fluid Meters, *Fluid Meters, Parts 1, 2, and 3*, ASME, New York, 1931–1937.
6. Tuve, G. L. and R. E. Sprenkle, *Instruments 6*, 201–206 (Nov. 1933).
7. McAdams, W. H., *Heat Transmission*, 2nd ed., McGraw-Hill, New York, 1942, p. 106.
8. *Chemical Engineers' Handbook*, 3rd ed., J. H. Perry, ed., McGraw-Hill, New York, 1950, p. 1453.
9. Keenan, J. H., *Thermodynamics*, Wiley, New York, 1941, p. 137.
10. Binney, A. M. and M. W. Woods, *Proc. Inst. Mech. Eng. 138*, 260 (1938).
11. Hoffman, E. J., *The Concept of Energy: An Inquiry into Origins and Applications*, Ann Arbor Science, Ann Arbor MI, 1977.
12. Govier, G. W. and K. Aziz, *The Flow of Complex Mixtures in Pipes*, Van Nostrand Reinhold, New York, 1972.

CHAPTER 2

The Conversion between Heat and Pressure – Volume Work

The conversion of heat to work is the main purpose for power cycles and the attendant investigations of thermodynamics. At some point in a power cycle there is involved the expansion of a gas from a higher pressure and temperature to a lower pressure and temperature. This is the power generation step. In what is called a heat pump or a refrigeration cycle, just the opposite occurs—the conversion of work to heat, which in mechanical cycles involves the compression of a gas.

The gas constitutes the working fluid, and may be retained wholly in the gaseous region (as in the Joule cycle) or may enter the two-phase region to be condensed and revaporized (as in the Rankine cycle). Pressurization prior to expansion may be produced by compression of the gas or by pumping the condensed liquid to pressure, followed by the addition of heat. To complete the cycle, the introduction of heat is required along with the rejection of heat. The difference between heat added and heat rejected becomes the net energy generated, which, on a time basis, is the power generated. The efficiency is, of course, the net energy generated divided by the heat added.

In the usual embodiment the compressed gases are heated or the pumped condensate is heated, vaporized, and superheated. In turn, heat is rejected by further cooling the expanded gases or by the cooling, condensation, and supercooling of the condensate prior to recycling to pressure. In one way or another, these steps are basic to conventional power cycles.

In all instances, the relationships for expansion and/or compression are fundamental to determining the effectiveness of power cycles. These relationships have been previously developed and are usually expressed and applied in terms of isentropic change. A brief recapitulation is in order first, for both open and closed systems, in terms of the corresponding differential and integrated forms. Both adiabatic and nonadiabatic behavior are of concern, but with a special emphasis on the latter. This recapitulation will be followed by an examination of the various classic cycles and their limitations, notably the hypothetical Carnot cycle, the practical Rankine cycle, and the Joule cycle. Although the Carnot cycle is assumed to be the most efficient cycle for converting heat to work, it can be argued that it is a nonallowable representation and that, under certain circumstances, the Rankine efficiency can be made to appear greater than the Carnot efficiency. Whereas the Rankine cycle is embodied in central steam-power plants, the closed Joule cycle is less well known, although the open Joule cycle or Brayton cycle is the basis for gas-fired turbine generators.

Inasmuch as the concept of entropy is often used to analyze power cycles, a critique is provided about its use and misuse. Among other things, for a real or nonideal gas, it is demonstrated that entropy is not a state function of temperature and pressure, contrary to what ordinarily is implicitly assumed. In other words, speaking in general, entropy does not yield a perfect differential, and its evaluation should properly depend upon the path of variation between temperature and pressure.

The Stirling cycle and the Ericsson cycle are described in passing because they are of theoretical interest. Another cycle of theoretical interest is a variant of the Joule cycle using isothermal compression and isentropic expansion.

Special attention is given to *nonadiabatic* compression and expansion in terms of the Joule cycle, which involves both the compression and expansion of a gaseous working fluid. Nonadiabatic differential behavior is examined first in detail and then extended to multistage behavior with interstage heat transfer. The latter is an application to be presented in depth in Chapter 4, after reviewing nuances of the Joule cycle in Chapter 3.

It is to be emphasized that modified Joule cycles using differential or multistage nonadiabatic compression and expansion furnish a means to circumvent the hypothetical efficiency limitations imposed by the Carnot cycle. This may be accomplished by heat transfer from the working fluid to a circulating or recirculating medium in a compressor section, with a combustive energy source or thermal energy source added to the medium after the compressor section, followed by heat transfer from the medium to the working fluid in an expander section. The medium may first be combustive air, either with fuel premixed or added after the compressor section. Ignition occurs after the compressor section and prior to the expander section. Alternately, the circulating medium may be a thermal or geothermal fluid, with fresh fluid added or injected after the compressor section and partially rejected after the expander section. In this manner, waste-heat from process heat exchangers becomes a low-grade thermal energy source. Even the waste-heat from the cooler–condensers used in the Rankine cycle at steam-power plants becomes a low-grade thermal energy source. These aspects are further detailed in Chapter 4.

Another cycle described at some length is a variation of the Rankine cycle known as the adjustable proportion fluid mixture (APFM) cycle, which sometimes is called the Kalina cycle. The APFM cycle involves a two-component working fluid rather than a single component as in the conventional Rankine cycle.

This chapter is rounded out by a discussion of heat pump and refrigeration cycles.

2.1. ISENTROPIC COMPRESSION AND EXPANSION

For an open system, in consistent units, as previously stated,

$$dH = C_p \, dT - C_p \mu \, dP = V \, dP + d\omega.$$

For the isentropic change of a perfect gas, $d\omega = d\text{lw} = T \, dS = 0$ and $\mu = 0$. Hence

$$C_p \, dT = V \, dP = \frac{RT}{P} \, dP,$$

which integrates to the customary expression

$$\frac{T_2}{T_1} = \left(\frac{P_2}{P_1}\right)^{R/C_p} = \left(\frac{P_2}{P_1}\right)^{(k-1)/k} = \alpha,$$

where $C_p - C_v = R$ such that $k = C_p/C_v$. Furthermore, for an adiabatic change, in consistent units (heat units),

$$-W = -\Delta W = C_p(T_2 - T_1) = C_p T_1(\alpha - 1) = C_p T_2\left(1 - \frac{1}{\alpha}\right).$$

The preceding expressions apply to either the compression or the expansion of a gas. For liquids, the specific volume V is a constant so that (in mechanical units)

$$-w = -\Delta w = V(P_2 - P_1).$$

The preceding relationship applies to either pumping or to a liquid-phase expansion.

For a closed system, in consistent units, as previously stated,

$$dU = C_v \, dT - C_v \eta \, dV = -P \, dV + d\omega.$$

For the isentropic change of a perfect gas, as before, $d\omega = d\text{lw} = T \, dS = 0$ and $\eta = 0$. Thus

$$C_v \, dT = -P \, dV = -\frac{RT}{V} \, dV,$$

which integrates to

$$\frac{T_2}{T_1} = \left(\frac{V_2}{V_1}\right)^{-R/C_v} = \left(\frac{V_2}{V_1}\right)^{1-k} = \left(\frac{T_2 P_1}{T_1 P_2}\right)^{1-k}$$

from which

$$\frac{T_2}{T_1} = \left(\frac{P_2}{P_1}\right)^{(k-1)/k} = \alpha.$$

The same identical result is obtained as for a flow system. In addition, for an adiabatic change (in heat units),

$$-W = -\Delta W = C_v(T_2 - T_1) = C_v T_1(\alpha - 1) = C_v T_2\left(1 - \frac{1}{\alpha}\right).$$

For liquids, if the specific volume V is constant, then the work requirement for pumping or a liquid-phase expansion is given (in mechanical units) by

$$-w = -\Delta w = P(V_2 - V_1) = 0;$$

that is, pumping or expansion is disallowed for an incompressible fluid in a closed system.

2.2. DIFFERENTIAL COMPRESSION AND EXPANSION

Nonadiabatic compression and expansion are ordinarily treated as discrete isentropic operations, where the rejection or addition of heat is made between stages. The subject of interstage cooling and heating will be dealt with extensively in Chapter 4 and may be regarded as the more practical method. Nevertheless, there is the occasion to consider further the integration or solution of the differential forms as introduced in Section 1.8 and as applied particularly to compression or expansion in the two-phase saturated region, e.g., to "wet compression."

Thus consider again the relationship

$$-dw = dH - dq = V\,dP + (d\omega - dq),$$

where for reversible nonadiabatic behavior the generalized irreversibility is zero,

$$d\omega^* = d\omega - dq = 0.$$

For a perfect gas it follows that

$$-dw = C_p\, dT - dq = \frac{RT}{P}\, dP$$

$$= \left(C_p - \frac{dq}{dT}\right) dT = \frac{RT}{P}\, dP,$$

where the derivative dq/dT may be made some arbitrary function of T or a constant; that is, as before, $dq/dT = f = f(T)$. Otherwise, some other form or condition of behavior may be prescribed.

For instance, as an alternative expression, it may be written that

$$-dw = C_p\, dT - \frac{dq}{dP}\, dP = \frac{RT}{P}\, dP,$$

where the derivative dq/dP may be made some arbitrary function of P or a constant. This representation is of utility in the consideration of isothermal behavior.

Solutions or integrations of the foregoing differential forms will be of interest here for three assumed conditions: isothermal change, change at varying temperature, and change along the saturated vapor curve. In all three cases the working fluid will exist only as a gas or vapor, albeit a saturated vapor in the last instance. The subject of compression and expansion in the two-phase region will again be taken up in Chapter 6.

Equivalency between Heat and Work for Isothermal Changes

For an open or flow system, as before,

$$dq - dw = dH = C_p\, dT - C_p \mu\, dP.$$

For an isothermal change, $dT = 0$, and for a perfect gas, $\mu = 0$. Therefore,

$$dq = dw \quad \text{and} \quad q = w \quad \text{or} \quad \Delta q = \Delta w.$$

The heat added must be equal to the work done.

For a closed system,

$$dq - dw = dU = C_v\, dT - C_v \eta\, dV.$$

For an isothermal change, $dT = 0$, and for a perfect gas, $\eta = 0$. Therefore, again,

$$dq = dw \quad \text{and} \quad q = w \quad \text{or} \quad \Delta q = \Delta w.$$

These idealizations may be used to evaluate energy cycles such as the Carnot cycle.

Note that for nonideal flow systems,

$$dq - dw = -C_p \mu \, dP,$$

and for nonideal closed systems,

$$dq - dw = -C_v \eta \, dV.$$

Moreover, in evaluating the pressure–volume integrals, the compressibility factor z may be introduced. Thus, for flow systems,

$$V \, dP = \frac{zRT}{P} \, dP,$$

and for closed systems,

$$-P \, dV = \frac{zRT}{V} \, dV.$$

The preceding relationships will have an effect on the behavior of real gases during compression and expansion. In particular, for nonideal gases during compression and expansion, $dw \neq dq$.

Isothermal Changes

For irreversible nonadiabatic isothermal behavior at a temperature T_0,

$$-w = -\Delta w = RT_0 \ln \frac{P_2}{P_1} + (\Delta \omega - \Delta q)$$

$$= RT_0 \ln \frac{P_2}{P_1} + \Delta \omega^*.$$

In an isothermal compression, heat is removed and Δq is negative. Assuming $\Delta \omega \sim 0$, then $\Delta \omega^*$ would be positive. The actual work requirement $(-\Delta w)$ would be greater than predicted by the first term on the right.

In an isothermal expansion from say point 1 to point 2, heat is added and Δq is positive. Again assuming $\Delta \omega \sim 0$, then $\Delta \omega^*$ would be negative. The actual work done (Δw) would be greater than predicted by the first term on the right; that is,

$$w = \Delta w = -RT_0 \ln \frac{P_2}{P_1} - \Delta \omega^*.$$

If reversible nonadiabatic behavior is assumed ($0 = d\omega^* = d\omega - dq$), then for an isothermal change in a flow system at some temperature T_0, an integration will yield

$$-w = -\Delta w = RT_0 \ln \frac{P_2}{P_1},$$

where it is implied that

$$q = \Delta q = \Delta \omega.$$

The working fluid irreversibilities ($\Delta \omega$) would equal the heat added to the working fluid. For isothermal behavior during compression, the heat added is negative and the working fluid irreversibilities also would be negative. During expansion, the heat added is positive and the working fluid irreversibilities would be positive.

Furthermore, according to the preceding section, for a reversible nonadiabatic isothermal compression or expansion, it is required that

$$q = \Delta q = \Delta \omega = w = \Delta w;$$

that is, the heat added is equal to the work done is equal to the working fluid irreversibilities per se.

For comparison, in an isentropic and adiabatic change, consider

$$-w = \int V \, dP = \int \frac{RT}{P} \, dP.$$

In compression, say, starting at T_0, since $T > T_0$, there will be a higher work input ($-w$) using an isentropic adiabatic change for the same specified pressure change or pressure ratio. In an expansion, say, starting at T_0, since $T < T_0$, there will be a lower work output using an isentropic adiabatic change for the same specified pressure change or pressure ratio.

Nonadiabatic Behavior at Varying Temperature

If cooling or heating occurs during differential compression or expansion, then, say, for a perfect gas it may be written that, as before,

$$-dw = (C_p - f) \, dT = \frac{RT}{P} \, dP.$$

These two equations constitute the statement of the problem.

If f is a constant between limits 1 and 2, then

$$\frac{T_2}{T_1} = \frac{P_2}{P_1}^{R/(C_p-f)}$$

and

$$-w = (C_p - f)(T_2 - T_1).$$

These expressions are of the same form as for an isentropic adiabatic change assuming C_p to be constant. The integrated relationships will apply to either compression or expansion, using the sign conventions for energy input and output. Note, however, that additionally

$$q = (f)(T_2 - T_1),$$

whereas in an isentropic adiabatic change, $q = 0$.

Effect of Heat Removal and Addition during Compression and Expansion

If the previous expressions for nonadiabatic differential behavior are integrated, then certain conclusions can be reached, as follows.

Compression. Integrating in the direction of compression, let

$$(-\Delta\overline{w})_{\text{comp}} = \int_{P_1}^{P_2} \frac{RT}{P}\, dP = \int_{T_1}^{T_2} C_p\, dT$$

$$-\int_1^2 d\overline{q}_{\text{comp}} = \int_{T_1}^{T_2} (C_p - f_{\text{comp}})\, dT,$$

where

$$f_{\text{comp}} = \left(\frac{d\overline{q}}{dT}\right)_{\text{comp}}.$$

Here, dT is positive for compression and, if heat is removed, $d\overline{q}$ is negative. Thus f_{comp} will be perceived as negative for the integration. Furthermore, at any given pressure, T will remain lower in value than during an isentropic adiabatic compression, starting from the same initial conditions of temperature and pressure; that is, the curve in P-T space will be steeper.

For a given or specified compression ratio, therefore, the result is that $-\Delta\bar{w}$ will have a lower positive value than for an isentropic adiabatic compression and T_2 will have a lower value also since the integrand will be larger; that is, the positive difference $(T_2 - T_1)$ will be smaller than for an isentropic adiabatic expansion.

Conceivably, the rate or degree of heat removal could be such that $T_2 \leq T_1$. Observe that when $T_2 = T_1$, which is the condition for isothermal compression,

$$(-\Delta\bar{w})_{\text{comp}} = \int_{P_1}^{P_2} \frac{RT}{P}\, dP = \int_1^2 (-d\bar{q})_{\text{comp}} = -(\bar{q}_2 - \bar{q}_1),$$

where $\bar{q}_1 = 0$ by definition. The work of compression would equal the heat lost.

Expansion. Integrating in the direction of expansion, let

$$(-\Delta\bar{w})_{\text{exp}} = \int_{P_3}^{P_4} \frac{RT}{P}\, dP = \int_{T_3}^{T_4} C_p\, dT$$

$$-\int_3^4 d\bar{q}_{\text{exp}} = \int_{T_3}^{T_4} (C_p - f_{\text{exp}})\, dT$$

or

$$\Delta\bar{w}_{\text{exp}} = -\int_{P_3}^{P_4} \frac{RT}{P}\, dP = -\left[\int_{T_3}^{T_4} C_p\, dT - \int_3^4 d\bar{q}_{\text{exp}} \right]$$

$$= -\int_{T_3}^{T_4} (C_p - f_{\text{exp}})\, dT,$$

where

$$f_{\text{exp}} = \left(\frac{d\bar{q}}{dT} \right)_{\text{exp}}.$$

Here, dT is negative and, if heat is added, dq is positive. Thus f_{exp} will again be perceived as negative for the integration. Furthermore, T will remain higher than during an isentropic adiabatic expansion, starting from the same initial conditions; that is, the curve in P-T space will appear to be steeper.

For a given expansion ratio, therefore, the result is that the positive value obtained for $\Delta\bar{w}$ will be greater than for an isentropic adiabatic expansion, as will the value of T_4; that is, the positive difference $T_3 - T_4$ will be smaller than for an isentropic adiabatic expansion.

As in the case of compression, if $T_4 = T_3$, which is the isothermal condition,

$$(-\Delta\bar{w})_{\exp} = \int_{P_3}^{P_4} \frac{RT}{P}\, dP = \int_3^4 (-d\bar{q}_{\exp}) = -(\bar{q}_4 - \bar{q}_3),$$

where $\bar{q}_3 = 0$. The heat added would equal the work done.

Locus of Temperature versus Pressure. The behavior of temperature versus pressure during compression and expansion is of interest. For compression, on separating the variables,

$$\int_{P_1}^{P_2} R\frac{dP}{P} = R\ln\frac{P_2}{P_1} = \int_{T_1}^{T_2} \frac{C_p - f_{\text{comp}}}{T}\, dT$$

is obtained. If f_{comp} is a constant, this reduces to the familiar forms

$$\ln\frac{P_2}{P_1} = \frac{C_p - f_{\text{comp}}}{R}\ln\left(\frac{T_2}{T_1}\right),$$

$$\ln\frac{T_2}{T_1} = \frac{R}{C_p - f_{\text{comp}}}\ln\left(\frac{P_2}{P_1}\right),$$

$$\frac{T_2}{T_1} = \left(\frac{P_2}{P_1}\right)^{R/C_p - f_{\text{comp}}}.$$

If $R = C_p - C_v$, then the exponent reduces to

$$\frac{R}{C_p - f_{\text{comp}}} = \frac{C_p/C_v - 1}{C_p/C_v - f_{\text{comp}}/C_v} = \frac{k - 1}{k - f_{\text{comp}}/C_v},$$

where f_{comp} is negative. If $f_{\text{comp}} = 0$, then the relationship for isentropic adiabatic compression is obtained.

Similarly for expansion, integrating in the direction of flow,

$$\int_{P_3}^{P_4} R\frac{dP}{P} = R\ln\frac{P_4}{P_3} = \int_{T_3}^{T_4} \frac{C_p - f_{\exp}}{T}\, dT.$$

If $f_{\exp}$ is constant, then the familiar form or forms is obtained as follows, where $P_4/P_3 < 1$ and $T_4/T_3 < 1$:

$$\ln\frac{P_4}{P_3} = \frac{C_p - f_{\exp}}{R}\ln\left(\frac{T_4}{T_3}\right)$$

$$\ln\frac{T_4}{T_3} = \frac{R}{C_p - f_{\exp}}\ln\left(\frac{P_4}{P_3}\right)$$

$$\frac{T_4}{T_3} = \left(\frac{P_4}{P_3}\right)^{R/C_p - f_{\exp}} .$$

where $f_{\exp}$ is again negative. The exponent may be transformed as for compression, and if $f_{\exp} = 0$, then the relationship for an isentropic adiabatic expansion is obtained.

Of further interest is the comparison that can be made for, say, $P_2/P_1 = P_3/P_4$, which has the connotations of a cycle; that is, $P_2 = P_3 = P_B$ and $P_1 = P_4 = P_A$, where P_B and P_A are the upper and lower pressure levels. Here it follows that

$$\int_{T_1}^{T_2}\frac{C_p - f_{\text{comp}}}{T}\,dT = \int_{T_4}^{T_3}\frac{C_p - f_{\exp}}{T}\,dT .$$

This can be perceived as the relationship between the temperature levels of a cycle and the behavior of f_{comp} and $f_{\exp}$ or vice versa.

If f_{comp} and $f_{\exp}$ are constants or mean values, then

$$(C_p - f_{\text{comp}})\ln\frac{T_2}{T_1} = (C_p - f_{\exp})\ln\frac{T_3}{T_4} .$$

If $T_3/T_4 > T_2/T_1$, then $(C_p - f_{\text{comp}}) > (C_p - f_{\exp})$, and vice versa. Alternatively, if $T_3/T_4 = T_2/T_1$, then $C_p - f_{\text{comp}} = C_p - f_{\exp}$. Recall that f_{comp} and $f_{\exp}$ are both negative, although they will, as expected, have different values.

Of special interest is the situation where $T_1 = T_4$. Here, if $T_3 > T_2$, then let $(C_p - f_{\text{comp}}) > (C_p - f_{\exp})$. That is to say, $-f_{\text{comp}} > -f_{\exp}$ and the slope of the compression curve will be greater than that of the expansion curve in P-T space, and similarly in $\ln P$-$\ln T$ space. Thus if heat is added between T_2 and T_3, the cycle still may be operated without heat rejection between T_4 and T_1. It is merely required that more total heat be added to the cycle (including that added during expansion) than is rejected during compression.

Comparison with Adiabatic Behavior. The *P-T* loci for nonadiabatic changes may be compared to an adiabatic change utilizing the following logarithmic forms:

Nonadiabatic Compression

$$\frac{d \ln P}{d \ln T} = C_p - f_{comp},$$

where f_{comp} is negative.

Adiabatic Change

$$\frac{d \ln P}{d \ln T} = C_p.$$

Nonadiabatic Expansion

$$\frac{d \ln P}{d \ln T} = C_p - f_{exp},$$

where f_{exp} is negative.

Comparison

The preceding derivatives indicate that in *P-T* space, starting at a given point, the nonadiabatic behavior for compression and expansion will each have a "slope" (dP/dT) at the same point in *P-T* space that is greater than for adiabatic behavior. Furthermore, the slope for nonadiabatic compression with heat rejected will be greater than the slope for nonadiabatic expansion if the compression and expansion legs are to be combined into a cycle. A similar situation exists in ln *P*-ln *T* space or in log *T*-log *P* space. In terms of the derivatives, the order is

$$\left(\frac{dP}{dT} \right)_{comp} > \left(\frac{dP}{dT} \right)_{exp} > \left(\frac{dP}{dT} \right)_{adiabatic}$$

or

$$\left(\frac{dT}{dP} \right)_{comp} < \left(\frac{dT}{dP} \right)_{exp} < \left(\frac{dT}{dP} \right)_{adiabatic}.$$

The net result is that during a nonadiabatic compression with heat removal, the working fluid temperature increases less than for a corresponding adiabatic compression that starts at the same point or the same initial condition. Consequently, less work is required for a given pressure change or compression ratio; that is, the integral of $-dw = V\,dP = (RT/P)\,dP$ will be smaller than for an adiabatic isentropic compression.

During a nonadiabatic expansion with heat added, the working fluid temperature decreases less than for a corresponding adiabatic expansion that starts at the same point or the same initial condition. Consequently, more work is produced for a given pressure drop or expansion ratio; that is, the integral of $dw = -V\,dP = -(RT/P)\,dP$ will be greater than for an adiabatic isentropic expansion.

These effects are built into the mathematics, whereby, for compression and expansion,

$$-\Delta\bar{w}_{comp} = \int_{P_1}^{P_2} \frac{RT}{P}\,dP = \left[\int_{T_1}^{T_2} C_p\,dT - \int_1^2 d\bar{q}_{comp}\right]$$

$$= \int_{T_1}^{T_2}(C_p - f_{comp})\,dT,$$

$$\Delta\bar{w}_{exp} = -\int_{P_3}^{P_4} \frac{RT}{P}\,dP = -\left[\int_{T_3}^{T_4} C_p\,dT - \int_3^4 d\bar{q}_{exp}\right]$$

$$= -\int_{T_3}^{T_4}(C_p - f_{exp})\,dT.$$

It turns out that $-\Delta\bar{w}_{comp}$ as in the preceding calculation will be smaller than for adiabatic behavior, for a given pressure change that starts at T_1. At the same time, $(d\ln P)/(d\ln T)$ will be larger than for adiabatic change or $(d\ln T)/(d\ln P)$ will be smaller. Another way of looking at it is that in the integral on the right of the equation, the integrand $C_p - f_{comp}$ is larger and, therefore, the temperature change or decrease must be smaller and the definite integral will be smaller.

It turns out that $\Delta\bar{w}_{exp}$ in the foregoing calculation will be greater than for adiabatic behavior, for a given pressure change that starts at T_3. At the same time, $(d\ln P)/(d\ln T)$ will be larger than for adiabatic change or $(d\ln T)/(d\ln P)$ will be smaller. Another way of looking at it is that in the integral on the right of the equation, the integrand $C_p - f_{exp}$ is larger and, therefore, the temperature change or decrease must be smaller, although the definite integral is larger.

Schematic comparisons of the P-T behavior and $\log T$-$\log P$ behavior are shown in Fig. 2.1 for the case where $T_4 > T_1$ and for the case where

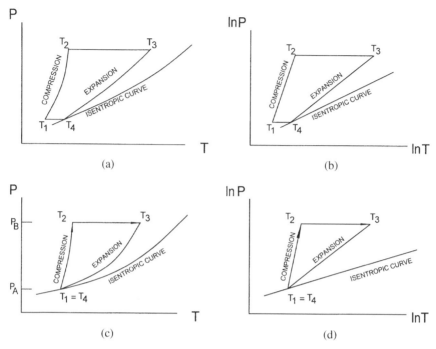

Fig. 2.1. Nonadiabatic pressure–temperature behavior. (a) $T_4 > T_1$ in P-T coordinates; (b) $T_4 > T_1$ in logarithmic coordinates; (c) $T_4 = T_1$ in P-T coordinates; (d) $T_4 = T_1$ in logarithmic coordinates.

$T_1 = T_4$. The relative juxtaposition is given for the adiabatic isentropic curve passing through T_4. The embodiment is for a cycle of combined compression and expansion, as will be discussed next.

Combined Compression and Expansion. The net work produced by a cycle composed of nonadiabatic compression and expansion can be stated as

$$\Delta \overline{w}_{\text{net}} = \Delta \overline{w}_{\text{comp}} + \Delta \overline{w}_{\text{exp}}$$

$$= -\int_{T_1}^{T_2}(C_p - f_{\text{comp}})\, dT - \int_{T_3}^{T_4}(C_p - f_{\text{exp}})\, dT,$$

where both f_{comp} and f_{exp} must be negative. If f_{comp} and f_{exp} have constant or mean values, then

$$\Delta \overline{w}_{\text{net}} = -(C_p - f_{\text{comp}})(T_2 - T_1) + (C_p - f_{\text{exp}})(T_3 - T_4),$$

where it will be required that

$$(C_p - f_{exp})(T_3 - T_4) > (C_p - f_{comp})(T_2 - T_1);$$

that is, the work done by the expansion must be greater than the work requirement for compression.

For the same compression and expansion pressure ratio $P_B/P_A = P_2/P_1 = P_3/P_4$,

$$(C_p - f_{comp})\ln\left(\frac{T_2}{T_1}\right) = (C_p - f_{exp})\ln\left(\frac{T_3}{T_4}\right).$$

Substituting for $(C_p - f_{exp})$, in the expression for $\Delta\overline{w}_{net}$ and rearranging yields

$$\Delta\overline{w}_{net} = (C_p - f_{comp})\ln\left(\frac{T_2}{T_1}\right)\left[-\frac{(T_2 - T_1)}{\ln(T_2/T_1)} + \frac{(T_3 - T_4)}{\ln(T_3/T_4)}\right].$$

Observe that if $T_1 = T_4$ and if $T_3 > T_2$, then $\Delta\overline{w}_{net}$ is positive. This situation corresponds to the case where heat is added between T_2 and T_3, but no heat is rejected between T_4 and T_1.

The subject is further discussed in Section 2.7 for differential behavior and in Chapter 4 for discrete behavior.

Special Case: Variable f_{comp} and f_{exp}. In the main, the function $f = dq/dT$ has been treated as if it were a constant. In general, however, f_{comp} and f_{exp} may be treated as variables, whose behavior, if predetermined, will ensure that a positive value is produced for the net work done by the cycle.

While the case has been considered where $T_1 = T_4$, there also may occur the case where $T_2 = T_3$ or where both situations coexist, that is, $T_1 = T_4$ and $T_2 = T_3$. It merely must be stipulated that the behavior of the functions f_{comp} and f_{exp} be such that the integrands $(C_p - f_{comp})$ and $(C_p - f_{exp})$, when integrated, respectively, in terms of T_{comp} and T_{exp} between the prescribed limits, give a positive value for the net work done; that is,

$$\Delta\overline{w}_{exp} > -\Delta\overline{w}_{comp}.$$

(It may be observed that T_{comp} and T_{exp} are different variables for the respective integrations, so that there is no stipulation that, say, $dT_{comp} = dT_{exp}$. In fact, as far as the mechanics of integration operations are concerned, it is immaterial what symbol is used for the variable of integration. Only the limits of integration are of interest. Any relationship may instead follow from the differential forms

$$(C_p - f_{comp}) \, dT_{comp} = \frac{RT_{comp}}{P_{comp}} \, dP_{comp},$$

$$(C_p - f_{exp}) \, dT_{exp} = \frac{RT_{exp}}{P_{exp}} \, dP_{exp},$$

whereby common operating pressure levels $P = P_{comp} = P_{exp}$ may be assumed, and if it can be assumed that $dP_{comp} = -dP_{exp}$ in the direction of flow and in the direction of the integrations, then

$$- \frac{dT_{exp}}{dT_{comp}} = \frac{T_{exp}}{T_{comp}} \left[\frac{C_p - f_{comp}}{C_p - f_{exp}} \right],$$

where the f-functions would be regarded as predetermined in terms of their respective temperatures.)

In P-T space, the locus of compression would be expected to exist to the left of the locus of expansion, such that, for any given pressure,

$$V \, dP_{comp} = \frac{RT_{comp}}{P_{comp}} \, dP_{comp}$$

in the direction of compression will be less than

$$- V \, dP_{exp} = - \frac{RT_{exp}}{P_{exp}} \, dP_{exp}$$

in the direction of expansion, where it may be stipulated that $dP_{comp} = -dP_{exp}$. The loci can be required to meet at, say, $T_1 = T_4$ and $T_2 = T_3$.

In effect, heat would be added only during expansion and rejected only during compression. Furthermore, more heat would be added than is rejected, which is evidenced by a positive value for the net work done.

Cycle Efficiency. A cycle efficiency can be defined as

$$\text{Eff} = \frac{\Delta \bar{w}_{net}}{\text{gross heat added}} \quad \text{or} \quad \text{Eff} = \frac{\Delta \bar{w}_{net}}{\text{net heat added}},$$

where again both f_{comp} and f_{exp} are to be negative. If f_{comp} and f_{exp} are constants or mean values, then, as before,

$$\Delta \bar{w}_{net} = -(C_p - f_{comp})(T_2 - T_1) + (C_p - f_{exp})(T_3 - T_4).$$

For the *gross* heat added to the working fluid, where, say, $T_1 = T_4$,

$$\text{gross heat added} = -f_{\exp}(T_3 - T_1) + C_p(T_3 - T_2)$$
$$= -f_{\exp}(T_3 - T_1) + C_p[(T_3 - T_1) - (T_2 - T_1)]$$
$$= -C_p(T_2 - T_1) + (C_p - f_{\exp})(T_3 - T_1)$$
$$= \Delta\bar{w}_{\text{net}} - f_{\text{comp}}(T_2 - T_1)$$

will be obtained. For the gross heat lost, considered as a positive quantity,

$$\text{gross heat lost} = -f_{\text{comp}}(T_2 - T_1) + C_p(T_4 - T_3).$$

Therefore, based on the gross heat added,

$$\text{Eff} = \frac{-C_p(T_2 - T_1) + (C_p - f_{\exp})(T_3 - T_4) + f_{\text{comp}}(T_2 - T_1)}{-C_p(T_2 - T_1) + (C_p - f_{\exp})(T_3 - T_4)}$$

$$= 1 - \frac{f_{\text{comp}}(T_2 - T_1)}{-C_p(T_2 - T_1) + (C_p - f_{\exp})(T_3 - T_4)}$$

$$= \frac{-(C_p - f_{\text{comp}})(T_2 - T_1) + (C_p - f_{\exp})(T_3 - T_4)}{-C_p(T_2 - T_1) + (C_p - f_{\exp})(T_3 - T_4)}.$$

For the *net* heat added to the working fluid,

net heat added
$$= \text{gross heat added minus gross heat lost}$$
$$= C_p(T_3 - T_2) + f_{\exp}(T_4 - T_3) - \left[-f_{\text{comp}}(T_2 - T_1) + C_p(T_4 - T_1)\right]$$
$$= f_{\text{comp}}(T_2 - T_1) - f_{\exp}(T_3 - T_4)$$
$$+ C_p[(T_3 - T_1) - (T_2 - T_1) - (T_4 - T_3) + (T_1 - T_3)]$$
$$= -(C_p - f_{\text{comp}})(T_2 - T_1) + (C_p - f_{\exp})(T_3 - T_4)$$
$$= \Delta\bar{w}_{\text{net}}$$

is obtained. However, as shown, this expression is identical to $\Delta\bar{w}_{\text{net}}$. The efficiency, therefore, at least hypothetically, could be 100%:

$$\text{Eff} = \frac{\Delta\bar{w}_{\text{net}}}{\Delta\bar{w}_{\text{net}}} = 100\%.$$

It is implicit in the preceding rating that the heat transferred between points 3 and 4 and the heat from compression are transferred to a medium, which is supplemented with or combined with a heat source, to raise the temperature of the working fluid from T_2 to T_3, and is further utilized as the heat source during expansion. Zero temperature differences between the working fluid and the medium may be assumed as the limiting case.

The usual comparison would be to the Carnot cycle efficiency, except that heat transfer becomes the criterion rather than temperature level. Furthermore, the heat exchange occurs at *varying* temperature rather than at constant temperature, as in the Carnot cycle.

Moreover, if the heat transferred during compression can be returned and utilized during expansion, as part of another heat source or sources, then the cycle efficiency can tend upward toward 100%, as demonstrated.

For instance, if heat transfer is made concurrently to a fluid medium during compression and can be re-utilized concurrently during expansion, along with another heat source, then with zero temperature approaches the efficiency could approach 100%. The situation can be viewed as if the Carnot cycle—or its equivalent—could be operated with recycle of the heat rejected. The subject is further explored in Section 2.7 and in Chapter 4.

Note that without the external heat source, the foregoing cycle would produce no net work and, discounting parasitic effects, it could at best only operate in perpetuity, sort of a closed-loop *perpetuum mobile*, going nowhere; that is, without the heat source, $T_2 = T_3$. The efficiency could be described as zero or at least as indeterminate. On the other hand, with the external source, an efficiency can be based on a rating for the external source. This rating for the external source can be the heat of combustion, say, or the sensible and latent heat changes which occur.

Cycle Efficiency with No Heat Recovery. If the heat removed during compression is not recovered, then as previously developed,

$$
\text{Eff} = \frac{\Delta \overline{W}_{\text{net}}}{-f_{\text{exp}}(T_3 - T_4) + C_p(T_3 - T_2)}
$$

$$
= \frac{-(C_p - f_{\text{comp}})(T_2 - T_1) + (C_p - f_{\text{exp}})(T_3 - T_4)}{-f_{\text{exp}}(T_3 - T_4) + C_p(T_3 - T_2)},
$$

where both f_{comp} and f_{exp} are negative. As a simplification, it can be assumed that $T_4 = T_1$.

Alternatively, in terms of a heat balance,

$$\text{Eff} = 1 - \frac{-f_{\text{comp}}(T_2 - T_1)}{-f_{\text{exp}}(T_3 - T_4) + C_p(T_3 - T_2)}.$$

The preceding two expressions are entirely equivalent.

In principle, a direct comparison cannot be made with Carnot efficiency since there are no constant temperature legs. However, as an approximation, let

$$\text{Eff}_{\text{Carnot}} \sim \frac{T_3 - T_1}{T_3} = 1 - \frac{T_1}{T_3}.$$

In the limit, the compression leg may be considered isothermal—a situation to be discussed in Section 2.7 under the heading "Isothermal Compression and Isentropic Expansion."

If the cycle efficiency is to be equal to the Carnot efficiency, then it is required that

$$\frac{-f_{\text{comp}}(T_2 - T_1)}{-f_{\text{exp}}(T_3 - T_4) + C_p(T_3 - T_2)} = \frac{T_1}{T_3}.$$

If the cycle efficiency is greater than the Carnot efficiency, then the term on the left will be smaller than the term on the right and vice versa.

In the foregoing expressions, if the expansion is adiabatic, then $f_{\text{exp}} = 0$. The effect will be to reduce the efficiency.

Limiting Conditions for Heat Removal and Addition

There are some limiting conditions or extenuating circumstances for the removal and addition of heat during differential compression and expansion which need to be further examined. Under consideration are the relationships for an ideal or perfect gas, which are expressed as

$$-dw = C_p \, dT - dq = V \, dP = \frac{RT}{P} \, dP$$

$$= (C_p - f) \, dT = V \, dP = \frac{RT}{P} \, dP,$$

where

$$f = \frac{dq}{dT}.$$

Alternatively,

$$-dw = C_p\, dT - \frac{dq}{dP} = V\, dP = \frac{RT}{P}\, dP,$$

which is of utility in considering isothermal behavior.

The function $f = dq/dT$ is normally negative for both nonadiabatic compression with heat removed and expansion with heat added, but may be zero for normal adiabatic isentropic behavior or may conceivably even change sign. As a special case, for isothermal behavior, it is required that dT or its derivatives be set equal to zero.

We are, therefore, interested in the following ranges:

$$(C_p - f) > C_p, \quad f \text{ is negative}, f < 0,$$

$$C_p > (C_p - f) > 0, \quad f \text{ is positive}, C_p > f > 0,$$

$$(C_p - f) < 0, \quad f \text{ is positive}, f > C_p.$$

There are in addition, as noted, the adiabatic isentropic changes which occur for $f = 0$ and the isothermal changes which occur for the case where $dT = 0$.

Compression. The limiting behavior may be determined from an examination of the differential behavior of $(C_p\, dT - dq)$. Whenever this difference is a positive value, it signifies that $(-dw)$ is positive and work is done upon the system, e.g., the work of compression. In other words,

$$C_p > \frac{dq}{dT} = f_{\text{comp}}$$

or

$$(C_p - f_{\text{comp}}) > 0.$$

Furthermore, it is required that dP be positive, and it follows that dT is positive, which are the normal attributes for compression. If heat is removed, then dq is negative, and if dT is positive, then f is negative.

RANGE: $f_{\text{comp}} < 0$ (heat removed)

The preceding statement is to include as a limiting case the qualification for adiabatic isentropic compression, where $f_{\text{comp}} = 0$. The region will extend down to and include isothermal behavior, but will not extend beyond this for reasons to be shown.

Isothermal Behavior. As f_{comp} becomes increasingly negative, a condition of apparent indeterminacy will be reached; that is, dP/dT must also increase without limit and $dT/dP \downarrow 0$. This presumably would be the requirement for isothermal behavior, whereby

$$-dw = (C_p - f_{comp})\, dT = C_p\, dT - dq = \frac{RT}{P}\, dP,$$

where $dT = 0$, and

$$-\Delta w = RT \ln \frac{P_2}{P_1}, \quad \text{where } P_2 > P_1.$$

Unfortunately, however, since $dT = 0$, the curious result is obtained that $(-dw) = -dq$ or $dw = dq$, and the total work of compression would have to equal the heat removed. So if all the work of compression ends up as heat removed, what would be left over for the actual pressure–volume work of compression? By definition, however, the system is a perfect gas ($\mu = 0$) and the enthalpy function ($dH = C_p\, dT - C_p \mu\, dP$) cannot change with pressure—only with temperature. Therefore, also by definition, since the temperature must remain constant, there can be no change in enthalpy of the working fluid, even though the pressure changes.

Thus it can be argued that for isothermal behavior, the positive quantity $-f_{comp} = -dq/dT$ must increase without limit, since $dT \downarrow 0$ where dq does not. Hence a contradiction or incongruity exists and/or an indefinite or indeterminate quantity is produced in order to meet the specification that, on the one hand, the work of compression $(-dw)$ is a positive nonzero value, $-dw = V\, dP$, but on the other hand it is required that $dH = 0$.

The assumptions and simplifications used, therefore, can be judged to be inadequate for the task for isothermal behavior.

Moreover and more appropriately, for isothermal behavior, we should be speaking of dq/dP rather than dq/dT. Thus it could be written that, for a real gas,

$$-dw = C_p\, dT - C_p \mu\, dP - \frac{dq}{dP}\, dP = V\, dP = \frac{zRT}{P}\, dP,$$

where the Joule–Thomson coefficient μ is introduced and where z is the compressibility factor. At constant temperature, presumably the effect of nonadiabatic behavior could be incorporated. However, for adiabatic behavior, the integration or solution of

$$C_p \, dT = \left[\frac{zRT}{P} + C_p \mu \right] dP$$

would be more complicated, since there is not a separation of variables, and would require numerical procedures.

Note, furthermore, that

$$-dw = -C_p \mu \, dP - dq$$

or

$$dq - dw = dH_T = -C_p \mu \, dP,$$

which establishes the enthalpic change at constant temperature for a real or nonideal gas. Thus, in general, $dq \neq dw$.

Additionally, it can be established that, since

$$-C_p \mu \, dP - \frac{dq}{dP} \, dP = \frac{zRT}{P} \, dP,$$

then

$$-\frac{dq}{dP} = C_p \mu + \frac{zRT}{P}.$$

This latter expression would give the required rate of heat transfer with pressure in order to maintain a constant temperature, say, for $T = T_1 - T_2$. For compression, dP is positive and dq is negative.

For an ideal gas, for isothermal behavior,

$$-\frac{dq}{dP} = -\frac{dw}{dP} = \frac{RT}{P},$$

which in consistent units would be equal to the unit volume or the reciprocal of the density.

Decreasing Temperature. If heat could be removed to an even greater degree than for isothermal change—that is, in the domain beyond isothermal behavior—the condition would be entered whereby dT/dP is negative; that is, the temperature would decrease with increasing pressure. This, however, would produce a contradiction, since $(C_p - f)$ remains positive, whereas dT/dP would become negative, which contradicts the energy balance.

RANGE: $0 < f_{\text{comp}} < C_p$ (heat added)

Proceeding in the other direction from adiabatic isentropic change, here again

$$(C_p \, dT - dq) > 0,$$

whereby

$$C_p > \frac{dq}{dT} = f_{\text{comp}}$$

or

$$(C_p - f_{\text{comp}}) > 0.$$

We are, however, interested in the situation for heat added; that is, dq is positive and dT remains positive. Thus as f_{comp} first becomes positive, where $(C_p - f_{\text{comp}})$ remains positive for $f_{\text{comp}} < C_p$, then for dP positive, dT is positive and $(-dw)$ is positive. Although the normal attributes of compression exist in terms of pressure and temperature and work required, heat would be added rather than removed.

Heat Transfer without Compression. At the point where

$$C_p \, dT - dq = 0$$

or

$$C_p = \frac{dq}{dT} = f_{\text{comp}}$$

a condition of indeterminancy is encountered whereby it would be required that $dP = 0$ or $dP/dT = 0$ and that $(-dw) = 0$, and there would be no work of compression. In other words, all the heat added would be going to increase the temperature of the gaseous working fluid. Heat transfer, only, would be involved.

RANGE: $f_{\text{comp}} > C_p$ (heat added)

Here the degree of heat added is extended such that

$$(C_p \, dT - dq) < 0$$

or

$$C_p < \frac{dq}{dT} = f_{comp}.$$

Thus if somehow f_{comp} should continue to become increasingly positive as heat is added, whereby $(C_p - f_{comp}) < 0$, then dP is negative and $(-dw)$ would be negative, or dw would be positive. The connotations for an expansion would exist. At least mathematically or hypothetically speaking, compression could yield to expansion, depending upon the degree of heat added. Therefore, if it is considered at all, heat addition during compression would have to be restricted to the former range where $0 < f_{comp} < C_p$, or oppositely stated, to the range where $C_p > dq/dT = f_{comp} > 0$.

CONCLUSIONS:

For reasons given in the preceding text, the degree of heat removal during compression will be restricted to the range from presumably isothermal behavior up to and including adiabatic isentropic behavior; that is, $f_{comp} \leq 0$. The degree of heat which might be added during compression will be restricted to the range between adiabatic isothermal behavior where $f_{comp} = 0$ and the case where $f_{comp} = C_p$; that is, $0 \preceq f_{comp} \preceq C_p$.

Expansion. The line of argument is similar, albeit inversely, to that for compression, and it is set forth as follows.

The limiting behavior here may also be determined from an examination of the behavior of $(C_p \, dT - dq)$. Whenever this difference is negative, it signifies that $(-dw)$ is negative or dw is positive, and work is done by the system, e.g., the work of expansion. In other words, if dT is negative,

$$C_p > \frac{dq}{dT} = f_{exp}$$

or

$$(C_p - f_{exp}) > 0.$$

Furthermore, it is required that dP be negative, and it follows that dT is also negative, which are the normal attributes for expansion. If heat is added, then dq is positive, and if dT is negative, then f is again negative, as is the case for compression. The relationships are mutually consistent.

RANGE: $f_{exp} < 0$ (heat added)

The statement must include as a limiting case the qualifications for adiabatic isentropic expansion, whereby $f_{exp} = 0$. The region will extend up to and including isothermal behavior, but will not extend beyond this for reasons to be shown.

Isothermal Behavior. As f_{exp} becomes increasingly negative, a condition of apparent indeterminacy will be reached; that is, dP/dT must also increase without limit or $dT/dP \downarrow 0$. This presumably would be the requirement for isothermal behavior, whereby

$$-dw = (C_p - f_{exp})\, dT = C_p\, dT - dq = \frac{RT}{P}\, dP,$$

where $dT = 0$, and

$$-\Delta w = RT \ln\left(\frac{P_4}{P_3}\right), \quad \text{where } P_4 < P_3.$$

Unfortunately, however, since $dT = 0$, then $-dw = -dq$ or $dw = dq$, and the total work of expansion would have to equal the heat added. So there is again the dilemma that on the one hand, $dH = 0$, and on the other hand, $-dw = V\,dP$.

Thus it can be argued that for the isothermal addition of heat during expansion, it is required that $-f_{exp} = -dq/dT$ increases without limit and that $(C_p - f_{exp})$ increase without limit, whereby $dT/dP \downarrow 0$.

More appropriately, therefore, we should be concerned with the behavior of dq/dP rather than with dq/dT, and we could introduce the Joule–Thomson coefficient as has been described for the case of compression. The derivative dq/dP will again be negative, since here dq is positive but dP is negative. The same considerations apply as for real gases versus ideal gases.

Increasing Temperature. The domain beyond isothermal behavior, where additional heat would be added, is excluded from consideration since a contradiction would occur; that is, $(C_p - f)$ remains positive, but dT/dP would have to become negative, which contradicts the energy balance.

RANGE: $0 < f_{exp} < C_p$ (heat removed)

Proceeding in the other direction from adiabatic isentropic change, but with heat removal, it is required that

$$(C_p\, dT - dq) < 0,$$

whereby, since dq is negative and dT must remain negative, then on dividing both sides of the equality by a negative quantity (dT), it must follow that the order is reversed such that

$$\left(C_p - \frac{dq}{dT} \right) dT < 0$$

becomes

$$\left(C_p - \frac{dq}{dT} \right) > 0.$$

Thus

$$C_p > \frac{dq}{dT} = f_{\text{exp}}$$

or

$$(C_p - f_{\text{exp}}) > 0.$$

The situation will be that if dq is negative and dP is negative, then dT remains negative and $(-dw)$ remains negative or dw is positive. The foregoing quantities continue as the normal attributes for expansion.

Thus as f_{exp} first becomes positive, where $(C_p - f_{\text{comp}})$ remains positive for $f_{\text{comp}} < C_p$, then for dP negative, dT is negative and $(-dw)$ is negative or dw is positive. Although the normal attributes of expansion exist in terms of pressure and temperature and work required, heat would be removed rather than added.

Heat Transfer without Expansion. At the point where

$$C_p \, dT - dq = 0$$

or

$$C_p = \frac{dq}{dT} = f_{\text{exp}},$$

a condition of indeterminancy is encountered, whereby it would be required that $dP = 0$ or $dP/dT = 0$ and that $(-dw) = 0$, and there would be no work of expansion. In other words, all the heat removed would be going to decrease the temperature of the gaseous working fluid. Heat transfer, only, would be involved.

RANGE: $f_{\text{exp}} > C_p$ (heat removed)

Here the degree of heat removal is extended such that

$$C_p < \frac{dq}{dT} = f_{\exp}.$$

Thus if somehow $f_{\exp}$ should continue to become increasingly positive as heat is removed, whereby $(C_p - f_{\exp}) < 0$, then since dT is negative, dP becomes positive and $(-dw)$ would be positive. The connotations for a compression would exist. At least mathematically or hypothetically speaking, expansion could yield to compression, depending upon the degree of heat removed.

Therefore, if it is considered at all, heat removal during expansion would have to be restricted to the range where $0 < f_{\text{comp}} < C_p$.

CONCLUSIONS:

In view of the foregoing considerations, the degree of heat addition during expansion will be restricted to that between isothermal behavior and isentropic adiabatic behavior, such that $f_{\exp} \leq 0$, and for heat removed, the range is $0 \leq f_{\exp} \leq C_p$.

Behavior of a Saturated Vapor

Here, the locus of compression or expansion will be defined by the vapor-pressure curve, which in the differential form may be expressed as

$$\frac{dP}{dT} = \frac{\Delta H_v}{T(V_g - V_L)} = \frac{P \Delta H_v}{RT^2},$$

where $\lambda = \Delta H_v$ is the latent heat of vaporization (a positive number) and is assumed to be constant over the interval. The liquid-phase specific or molar volume V_L is regarded as comparatively negligible, and the vapor-phase specific or molar volume V_g or V_v or V_V is assumed to be reproducible by the perfect gas law. Moreover, all units are to be consistent. Therefore, on substituting,

$$-d\overline{w} = \frac{RT}{P} \frac{P \Delta H_v}{RT^2} \, dT = \Delta H_v \, d \ln T$$

and

$$-w = \Delta H_v \ln \frac{T_2}{T_1}.$$

Also,

$$q = C_p(T_2 - T_1) - \Delta H_v \ln \frac{T_2}{T_1},$$

where, by the enthalpic energy balance for the fluid, $H_2 - H_1 = C_p(T_2 - T_1)$.

The preceding relationships would provide the work and heat requirements for a saturated vapor to follow the vapor-pressure curve during compression or expansion. There would by definition be no coexisting liquid phase; the working fluid would be 100% saturated vapor at the so-called dew-point condition.

Use of Pressure Ratios. From the integrated vapor-pressure relationship, between pressure levels P_A and P_B,

$$\frac{1}{T_2} - \frac{1}{T_1} = -\frac{R}{\Delta H_v} \ln \frac{P_B}{P_A}$$

or

$$\frac{T_2}{T_1} = 1 + \frac{RT_2}{\Delta H_v} \ln \frac{P_B}{P_A}$$

so that

$$-w = \Delta H_v \ln \left[1 + \frac{RT_2}{\Delta H_v} \ln \frac{P_B}{P_A} \right].$$

The temperatures T_1 and T_2 correspond to P_A and P_B, respectively. Alternately,

$$\frac{T_2}{T_1} = 1 + \frac{RT_1}{\Delta H_v} \ln \frac{P_B}{P_A}$$

so that

$$-w = \Delta H_v \ln \left[1 - \frac{RT_1}{\Delta H_v} \ln \frac{P_B}{P_A} \right].$$

More commonly, the use of the initial temperature T_1 for the reference will be preferable.

Heat Requirement. The heat requirement will be

$$q = C_p(T_2 - T_1) + w.$$

Whether the heat requirement will be positive or negative will depend upon the relative value of C_p versus $\Delta H/T$. Thus, in the differential form, consider

$$dq = C_p \, dT - \Delta H_v \, d \ln T = \left(C_p - \frac{\Delta H_v}{T} \right) dT.$$

For compression, for example, if $C_p < \Delta H_v/T$, then dq will be negative. Other things being equal, this condition is favored for lower temperature ranges, well below the critical, since the latent heat of vaporization $\lambda = \Delta H_v$ approaches zero at the critical point. There will be a crossover at $C_p = \Delta H_v/T$, whereby dq would become positive at higher temperatures, requiring the addition of heat to stay at the saturated vapor condition along the vapor-pressure curve.

For expansion, the conditions and requirements would be the converse.

Comparison with Isentropic Behavior. For an isentropic change, at a varying temperature T',

$$-dw = C_p \, dT' = \frac{RT'}{P} \, dP$$

from which

$$C_p d \ln T' = Rd \ln P$$

and

$$C_p \ln \frac{T_2'}{T_1} = R \ln \frac{P_B}{P_A},$$

where T_2' represents the final temperature level encountered at P_B and where $T_1 = T_1'$.

For a change along the vapor-pressure curve, at a varying temperature T, starting at $T_1 = T_1'$,

$$-dw = \frac{\Delta H_v}{T} \, dT = \frac{RT}{P} \, dP$$

from which

$$\frac{\Delta H_v}{T} d \ln T = R d \ln P$$

or

$$\Delta H_v \frac{dT}{T^2} = R d \ln P$$

and

$$-\Delta H_v \left(\frac{1}{T_2} - \frac{1}{T_1} \right) = R \ln \frac{P_B}{P_A}.$$

This expression merely reproduces the vapor-pressure curve.

It can be observed that, during compression, for instance, if $C_p <$ $\Delta H_v/T$, then for a given pressure increase, the temperature T reached during an isentropic adiabatic compression will be greater than that which would be reached during a compression along the vapor-pressure curve.

Furthermore, the work requirement for an isentropic adiabatic compression will be greater than for the compression along the vapor-pressure curve, for the same pressure change or compression ratio.

Although the comparison can be made, in principle, using the integrated relationships, it is much simpler to compare the differential forms. Thus let T' be the temperature during the isentropic adiabatic compression and let T be the temperature for the compression along the vapor-pressure curve. Therefore, from the energy balances for the same pressure change,

$$\Delta T' = \frac{\Delta H_v/T}{C_p} \frac{T'}{T} dT,$$

where it may be assumed that initially $T' = T$. If $C_p < \Delta H_v/T$, then $dT' > dT$; that is, the temperature will rise faster for the isentropic adiabatic compression than for the compression along the vapor-pressure curve.

Moreover, on rearranging the preceding expression,

$$C_p \, dT' = \frac{T'}{T} \Delta H_v \ln T$$

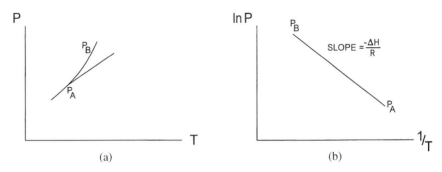

Fig. 2.2. Compression or expansion along the vapor-pressure curve.

or

$$-w_{\text{isentropic}} = \frac{T'}{T}(-w_{\text{vapor-pressure}}).$$

Since T'/T is greater than unity, then the isentropic compression work requirement will be greater than that for compression along the vapor-pressure curve, whenever $C_p < \Delta H_v/T$; that is, the isentropic compression path will tend to move away from the vapor-pressure curve and into the gaseous single-phase region. Another way of saying this is that the slope of the isentropic locus in *P-T* coordinates (pressure as the ordinate and temperature as the abscissa) will be less than the corresponding slope of the vapor-pressure curve.

A comparison of routes is shown in Fig. 2.2 in terms of both linear and semilog coordinates. In Fig. 2.2(a) the segment $P_A - P_B$ denotes change along the vapor-pressure curve relative to isentropic change. In Figure 2.2(b) this change appears as a straight line with negative slope.

Of special note is the fact that vapor-pressure curves also may be approximated as a straight line in log-log coordinates, that is, for $\log P$ versus $\log T$. Thus using the first approximation for the logarithm, over a suitable range of T,

$$\ln T \sim \frac{T-1}{T} = 1 - \frac{1}{T} = \ln e - \frac{1}{T},$$

whereby

$$-\frac{1}{T} \sim \ln T - \ln e.$$

From the Clausius–Clapeyron integrated form for the vapor-pressure relationship, where ΔH_v is the latent heat of vaporization,

$$
\begin{aligned}
\ln P &= \frac{-\Delta H_v}{RT} + c \\[2mm]
&= \frac{-\Delta H_v}{RT} + \ln c' \\[2mm]
&= \frac{\Delta H_v}{R}\ln T - \frac{\Delta H_v}{R}\ln e + \ln c' \\[2mm]
&= \frac{\Delta H_v}{R}\ln T + \ln\left[c' \exp\left(\frac{-\Delta H_v}{R}\right)\right],
\end{aligned}
$$

where the term in brackets is a constant. Thus a plot of log P versus log T may approximate a straight line in log-log coordinates.

EXAMPLE 2.1

A comparison is to be made for steam, for the following conditions and properties:

$(k - 1)/k$	0.23
T_1	460 + 212 = 672°R
P_A	1 atm
P_B	10 atm
ΔH_v	1000(18) = 18,000 Btu/lb-mol
C_p	0.5 Btu/lb-mol-°F
MW	18

Isentropic Adiabatic Compression

$$
\frac{T_2'}{T_1} = \left(\frac{10}{1}\right)^{0.23} = 1.698,
$$

$$
-w = C_p T_1\left[\left(\frac{P_B}{P_A}\right)^{0.23} - 1\right] = (0.5)(672)[1.698 - 1] = 234.5 \text{ Btu/lb,}
$$

$$
q = 0.
$$

Compression along the Vapor–Pressure Curve

$$\frac{T_2}{T_1} = \left\{ 1 - \left[\frac{(1.987)(672)}{18,000} \right] \ln 10 \right\}^{-1} = 1.2060,$$

$$-w = -1000 \ln(1.2060) = 187.3 \, \text{Btu/lb},$$

$$q = C_p(T_2 - T_1) + w = 0.5(672)(1.2060 - 1) - 187.3 = -118.1 \, \text{Btu/lb}.$$

Isothermal Compression

$$-w = RT \ln \frac{P_B}{P_A} = \left[\frac{1.987}{18} \right](672)\ln 10 = 170.8 \, \text{Btu/lb},$$

$$q = w = -170.8 \, \text{Btu/lb}.$$

2.3. THE CARNOT CYCLE

The classic reference cycle for the conversion of heat to pressure–volume work is the hypothetical Carnot cycle, as devised first in Carnot's *Memoir on the Motive Power of Fire*. We will subsequently examine why the cycle is theoretically impossible, although, nevertheless, useful as a reference standard.

The physical embodiment involves a cylinder with a piston where heat added produces an expansion and heat removed produces a contraction. The cycle is diagrammed in Figs. 2.3 and 2.4 for various coordinates. Leg 1-2 is an adiabatic compression to a higher temperature, followed by an isothermal expansion to a lower pressure. This is in turn followed by leg 3-4, an adiabatic expansion to a lower temperature. Leg 4-1, an isothermal compression, completes the cycle.

There will be more heat added than removed; the difference is presumed to be converted to pressure–volume work. The determination may be made based on either a closed or an open system. The former basis will be pursued here. With the adiabatic compression and expansion assumed isentropic, where $k = C_p/C_v$, the quantitative relationships are as follows:

Leg 1-2 (Adiabatic compression):

$$\frac{T_2}{T_1} = \left(\frac{V_2}{V_1} \right)^{1-k} = \alpha,$$

$$-\Delta w_{1\text{-}2} = J_0 C_v(T_2 - T_1) = J_0 C_v T_1(\alpha - 1).$$

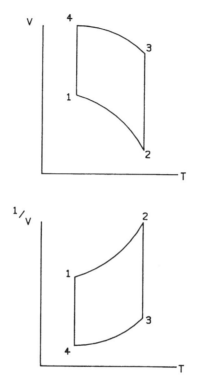

Fig. 2.3. Carnot cycle in *V-T* and 1/*V-T* coordinates.

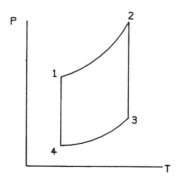

Fig. 2.4. Carnot cycle in *P-T* coordinates.

Leg 2-3 (Isothermal expansion, $T_2 = T_3$):

$$-\Delta w_{2\text{-}3} = -\Delta q_{2\text{-}3} = RT_2 \ln \frac{V_3}{V_2} + \Delta \omega_{2\text{-}3}^*.$$

Leg 3-4 (Adiabatic expansion):

$$\frac{T_3}{T_4} = \left(\frac{V_3}{V_4}\right)^{1-k} = \beta,$$

$$-\Delta w_{3\text{-}4} = J_0 C_v (T_4 - T_3) = J_0 C_v T_4 (1 - \beta) = J_0 C_v T_3 \frac{1}{\beta - 1}.$$

Leg 4-1 (Isothermal compression, $T_4 = T_1$):

$$-\Delta w_{4\text{-}1} = -\Delta q_{4\text{-}1} = RT_1 \ln \frac{V_1}{V_4} + \Delta \omega_{4\text{-}1}^*.$$

It will follow that

$$\frac{T_2}{T_1}\frac{T_4}{T_3} = \frac{T_2}{T_3}\frac{T_4}{T_1} = 1 = \left[\frac{V_2}{V_1}\frac{V_4}{V_3}\right]^{1-k}.$$

Therefore,

$$\frac{V_3}{V_4} = \frac{V_2}{V_1} \quad \text{and} \quad \frac{V_3}{V_2} = \frac{V_4}{V_1}.$$

Consequently, for the temperature ratios, from the former,

$$\alpha = \frac{T_2}{T_1} = \frac{T_3}{T_4} = \beta;$$

from the latter,

$$\ln \frac{V_3}{V_2} = \ln \frac{V_4}{V_1} = -\ln \frac{V_1}{V_4}.$$

From a heat balance, it may be stated that, for isothermal changes, where $\Delta q = \Delta w$,

$$\text{Eff}_{\text{Carnot}} = \frac{\Delta q_{2\text{-}3} + \Delta q_{4\text{-}1}}{\Delta q_{2\text{-}3}}$$

$$= \frac{-RT_2 \ln(V_3/V_2) - \Delta \omega^*_{2\text{-}3} - RT_1 \ln(V_1/V_4) - \Delta \omega^*_{4\text{-}1}}{-RT_2 \ln(V_3/V_2) - \Delta \omega^*_{2\text{-}3}}.$$

If $\Delta \omega^*_{2\text{-}3} = \Delta \omega^*_{4\text{-}1} = 0$, then it may be inferred that

$$\text{Eff}_{\text{Carnot}} = \frac{T_2 - T_1}{T_2} = \frac{T_h - T_c}{T_h},$$

where T_h and T_c denote the temperatures along the hot isothermal leg and cold isothermal leg, respectively.

There is also the inference that $-\Delta \omega^*_{2\text{-}3} - \Delta \omega^*_{4\text{-}1} = 0$, where $\Delta \omega^*_{2\text{-}3}$ for the isothermal expansion is negative and $\Delta \omega^*_{4\text{-}1}$ for the isothermal compression is positive. The fact that $\Delta \omega^*_{2\text{-}3}$ will be negative lowers the efficiency.

Note that although the sum of the generalized irreversibilities $\Delta \omega^*$ or $-\Delta \omega^*$ may tend to cancel, the efficiency may still decrease. There is the possibility, moreover, that with sufficiently high values for the irreversibilities, the efficiency may become zero—or even negative, which is not allowable. This is an indication, therefore, that the Joule–Thomson coefficient or Gay–Lussac coefficient should be taken into account for isothermal changes; that is, for a closed system at constant temperature,

$$dq - dw = dU = -C_v \eta \, dP = -P \, dV + d\omega$$

or

$$-dw = -C_v \eta \, dP - dq$$
$$= -P \, dV + (d\omega - dq)$$
$$= -P \, dV + d\omega^*$$

such that $dq \neq dw$. A similar representation may be made for a flow system in terms of the enthalpy and Joule–Thomson coefficient.

From a different standpoint, there can be no internal energy change for an ideal working fluid in a closed system during either an isothermal compression or expansion; that is,

$$\Delta q - \Delta w = 0 = \Delta U.$$

However, it is also required that

$$0 = \Delta U = -\int P\,dV + \Delta\omega.$$

If $\Delta\omega = 0$, then $\int P\,dV = 0$. This in itself is a contradiction, since a volume change occurs in both isothermal compression and expansion.

On the other hand, if

$$\Delta\omega = \int P\,dV,$$

then this in turn may produce a contradiction; that is, $\Delta\omega = \int P\,dV$ will take on a negative value during compression, whereas it can be argued that the intrinsic energy change or irreversibility $\Delta\omega$ should always be positive, at least in fluid flow. During expansion, this sort of discrepancy is not the case, since $\Delta\omega$ would be positive. A resolution requires that, in isothermal compression and expansion, $\Delta\omega$ be defined by the integral.

A similar situation exists for a flow system in isothermal compression and expansion, where

$$\Delta\omega = -\int V\,dP$$

is also negative for compression but positive for expansion.

Finally, the argument can be made that the foregoing notion is merely an exercise in circularity. Thus consider the isothermal expansion for leg 2-3, as before:

$$-\Delta w_{2\text{-}3} = -\Delta q_{2\text{-}3}$$

$$= RT_2 \ln\frac{V_3}{V_2} + \Delta\omega^*_{2\text{-}3},$$

where $\Delta\omega^*_{2\text{-}3} = \Delta\omega_{2\text{-}3} - \Delta q_{2\text{-}3}$ and $\Delta\omega_{2\text{-}3} = -RT_2 \ln(V_3/V_2)$. On substituting, all that is obtained is

$$-\Delta w_{2\text{-}3} = -\Delta q_{2\text{-}3} = -\Delta q_{2\text{-}3}.$$

Similar results are obtained for other isothermal compressions or expansions. In an attempt to be completely rigorous, only an identity is obtained.

Nonadiabatic Isentropic Behavior

The dilemma for isothermal compression and expansion can also be examined in terms of nonadiabatic isentropic changes. First assuming constant temperature, in the limit, the Gay–Lussac coefficient and the Joule–Thomson coefficient approach zero, producing contradictions.

Consider a closed system whereby

$$dq - dw = dU = C_v\, dT - C_v \eta\, dV = -P\, dV + d\omega,$$

where $d\omega = 0$ for isentropic behavior. If $dT = 0$, then

$$-C_v \eta\, dV = -P\, dV \quad \text{or} \quad C_v \eta = P.$$

Thus if $\eta \to 0$ for an ideal gas, then $P \to 0$, which is not allowable.
 For a flow system,

$$dq - dw = dH = C_p\, dT - C_p \mu\, dV = V\, dP + d\omega,$$

where $d\omega = 0$ for isentropic behavior. If $dT = 0$, then

$$-C_p \mu\, dP = V\, dP \quad \text{or} \quad -C_p \mu = V.$$

This is not only a contradiction in signs, since μ is most generally positive for gases, but for a perfect gas $\mu \to 0$, so that $V \to 0$, which is not allowable.

We may conclude, therefore, that neither isothermal compression nor isothermal expansion is an allowable phenomenon and, consequently, neither is the Carnot cycle. Alternatively, at least the attempt to represent these isothermal phenomena mathematically will run into physical (and mathematical) contradictions. In turn, the efficiency calculation for the Carnot cycle must be considered suspect.

2.4. THE RANKINE CYCLE

The Rankine cycle involves the phase changes which occur in a working fluid. The expansion of a vapor through a turbine or turbines occurs, followed by condensation of the exhaust vapors. The condensate is pumped back to the initial working pressure, vaporized, and superheated to the initial condition.

The ubiquitous working fluid is the water–steam system, although other fluids, such as mercury, ammonia, Freons, or hydrocarbons, can be used, albeit with altered specifications and efficiencies.

The Rankine cycle is diagrammed in Fig. 2.5 in P-T coordinates. Leg 3-4 may be idealized as an isentropic expansion; leg 4-1 represents cooling and condensation (and supercooling). There is a phase change at R. Leg 1-2 denotes pumping and leg 2-3 represents reheating, vaporization, and superheating. There is a phase change at S.

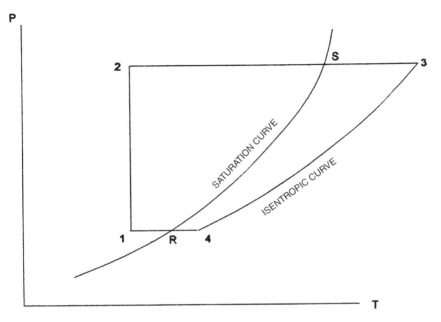

Fig. 2.5. Rankine cycle in *P-T* coordinates.

In terms of the enthalpy function H, an efficiency may be defined as

$$\text{Eff}_{\text{Rankine}} = \frac{(H_3 - H_4) - (-\Delta W_p)}{H_3 - H_2},$$

where $(-\Delta W_p)$ is positive and denotes parasitic losses, including pumping and circulating losses. Furthermore, to make the units consistent,

$$H_2 - H_1 = -\Delta W_p = \frac{-\Delta w_p}{J_0}.$$

Since enthalpy is a state function, and thus the behavior of the enthalpy functions and the enthalpy changes or differences is independent of path, then

$$H_3 - H_2 = (H_3 - H_4) + (H_4 - H_R) + \lambda_R + (H_R' - H_1) - (H_2 - H_1),$$

where λ_R denotes the latent heat of vaporization—a positive value—in consistent units. Furthermore, by definition, $H_R = H_R' + \lambda_R$; that is, if H_R is the enthalpy of the saturated vapor at equilibrium, then H_R' is the corresponding enthalpy of the saturated liquid.

A more customary relationship for the heat input at pressure level P_2 is

$$
\begin{aligned}
H_3 - H_2 &= (H_3 - H_S) + \lambda_S + (H_S' - H_2) \\
&= (C_p)_V|_{P_2}(T_3 - T_S) + \lambda_S + (C_p)_L|_{P_2}(T_S - T_2),
\end{aligned}
$$

where $(C_p)_V|_{P_2}$ represents the mean heat capacity of the vapor phase and $(C_p)_L|_{P_2}$ represents the mean heat capacity of the liquid phase. Also, $H_S = H_S' + \lambda_S$, where for some few practical purposes $\lambda_S \sim \lambda_R$.

Equating the two expressions for the heat input $H_3 - H_2$ and solving for $H_2 - H_1$, we find that

$$
\begin{aligned}
H_2 - H_1 &= (H_3 - H_4) + (H_4 - H_R) + \lambda_R + (H_R' - H_1) \\
&\quad -[(H_3 - H_S) + \lambda_S + (H_S' - H_2)] \\
&= C_p(T_3 - T_4) + (H_4 - H_R) + \lambda_R + (H_R' - H_1) \\
&\quad -\left[(C_p)_V|_{P_2}(T_3 - T_S) + \lambda_S + (C_p)_L|_{P_2}(T_S - T_2)\right].
\end{aligned}
$$

This can be construed as the rigorous expression for the pumping requirement per se, assuming adiabatic behavior, whereby we may write

$$
H_2 - H_1 = -\Delta W_{\text{pump}}.
$$

The pumping requirement between levels P_1 and P_2 may instead be approximated by

$$
H_2 - H_1 = -\Delta W_{\text{pump}} \sim \frac{1}{J_0}\frac{1}{\rho_L}(P_2 - P_1).
$$

Usually the fluid temperature is assumed constant. This expression may or may not be interpreted as including other dissipative circulating effects, which ordinarily are accommodated by the inclusion of an extra term or terms, with the entirety entered under parasitic effects. All in all, attempting to relate pumping requirements to fluid enthalpic changes is a complicated affair.

Some simplifications are in order.

If there is no cooling of the exhaust vapors—that is, the exhaust vapors reach the saturation or equilibrium curve—then as a simplification, $(H_4 -$

$H_R) = 0$. If there is no supercooling of the condensate, $(H_R' - H_1) = 0$. Under these circumstances, therefore,

$$\text{Eff}_{\text{Rankine}} = \frac{(H_3 - H_4) - (-\Delta W_p)}{(H_3 - H_4) + \lambda_R - (H_2 - H_1)}$$

$$= \frac{(H_3 - H_4) - (-\Delta W_p)}{(H_3 - H_4) + \lambda_R - (-\Delta W_{\text{pump}})}.$$

Furthermore, if $(C_p)_V|_{P_2} \sim C_p$ and $T_1 = T_2 = T_4 = T_R$, then

$$H_2 - H_1 = -\Delta W_{\text{pump}} \sim C_p(T_S - T_4) - (C_p)_L|_{P_2}(T_S - T_4) + (\lambda_R - \lambda_S)$$

$$\sim \left[C_p - (C_p)_L|_{P_2} \right](T_S - T_4) + (\lambda_R - \lambda_S).$$

This term by definition must be positive; that is, the latent heat difference more than offsets the sensible heat changes. If the pumping requirement is assumed negligible, this is equivalent to assuming that the latent heat difference is equal to the sensible heat changes. Inasmuch as mean values are assumed for the heat capacities, there is an element of rigor that is lost.

Neglecting the parasitic requirements as such, and assuming an ideal isentropic expansion,

$$\text{Eff}_{\text{Rankine}} = \frac{C_p(T_3 - T_4)}{C_p(T_3 - T_4) + \lambda_R - (-\Delta W_{\text{pump}})}$$

$$= 1 \Big/ \left(1 + \frac{\lambda_R/C_p}{T_3 - T_4} - \frac{-\Delta W_{\text{pump}}/C_p}{T_3 - T_4} \right)$$

$$= \frac{T_3 - T_4}{(T_3 - T_4) + \lambda_R/C_p - (-\Delta W_{\text{pump}})/C_p}$$

$$= \frac{T_3 - T_4}{T_3 - \left[T_4 - \lambda_R/C_p + (-\Delta W_{\text{pump}})/C_p \right]},$$

where $T_1 = T_R = T_4$, which is the saturation temperature of the condensate. Note that here the latent heat of vaporization pertains to the condensate condition.

The last-cited equation, bears a striking resemblance to the equation form for the Carnot efficiency. Observe, furthermore, that if

$$T_4 > \left[\frac{\lambda_R}{C_p} - (-\Delta W_{\text{pump}}) \right],$$

then the Rankine efficiency expression would give a greater value than the Carnot efficiency expression. In addition, if $T_4 = [\lambda_R/C_p - (-\Delta W_{\text{pump}})]$, then the efficiency expressions would give the same value, whereas if $T_4 < [\lambda_R/C_p - (-\Delta W_{\text{pump}})]$, then the Rankine efficiency value would be less than the Carnot efficiency value. In all cases, the convention is that $H_2 - H_1 = (-\Delta W_{\text{pump}})$.

Note, however, that the two cycles are essentially different from one another. For instance, in the Carnot cycle the higher temperature T_3 would be constant over a range of behavior, whereas in the Rankine cycle, T_3 is only the value at a point; similarly for the lower temperature T_4, which pertains to a range of behavior in the Carnot cycle, but which may pertain only to a point (the point of condensation) in the Rankine cycle. Thus to try to make a comparison is like comparing "apples and oranges." The two cycles are simply different, with no commonality for a comparison.

The Subject of Antinomies. An "antinomy" is defined as the contradiction between two principles or conclusions, each taken to be correct or true. Thus if one point of view asserts that what we refer to as the Carnot cycle is *always* the most efficient way to convert, yet another point of view shows that this is not always the case, then we have an *antinomy*.

The very fact that there is such a technical word as "antinomy" in logical or philosophical circles indicates that the foregoing circumstance is by no means the first time that a logical contradiction has been incurred.

Subcritical Region. The previously derived expressions may be examined in the subcritical region—the region below the critical point. Thus consider again

$$\text{Eff}_{\text{Rankine}} = \frac{T_3 - T_4}{T_3 - T_4 + \lambda_R/C_P - (-\Delta W_{\text{pump}})/C_p},$$

where $(-\Delta W_{\text{pump}})$ will have a positive value. As the equilibrium curve or saturation curve or vapor-pressure curve approaches the critical point, the value for the latent heat of vaporization diminishes and becomes zero in the limit at the critical point. The temperature difference $T_3 - T_4$ will expectedly diminish also, however. To illustrate the combined effect, the

steam tables may be consulted. Thus let T_3 denote the critical temperature, whereby $T_3 = 705.4°F$ or $1165.4°R$. The critical pressure is 3206.2
psia. Arbitrarily choose $T_4 = T_R = 700°F$ or $1160°R$, at which point $\lambda_R = 172.1$ Btu/lb. The corresponding saturation pressure will be 3093.7 psia.
Furthermore, for the vapor or gaseous phase at circa these conditions,
$C_p \sim 3.6$ Btu/lb-°F or 3.6 Btu/lb-°R.

Neglecting the pumping requirement, for the Rankine efficiency,

$$\text{Eff}_{\text{Rankine}} = \frac{5.4}{5.4 + 172.1/3.6} = 0.1015 = 10.15\%.$$

Note that if the pumping requirement was included in the denominator,
the Rankine efficiency would increase, whereas, for the Carnot efficiency,

$$\text{Eff}_{\text{Carnot}} = \frac{T_3 - T_4}{T_3} = \frac{5.4}{1165.4} = 0.0046 = 0.46\%;$$

that is, for small differences between the operating temperature levels
immediately below the critical, the Carnot efficiency would become small,
or very small, and here would be about 20 times lower than the ideal
calculated Rankine efficiency. The discrepancy would be even greater
if the pumping requirement was included in the Rankine efficiency
calculation.

Heat Balance Closure

The enthalpy difference for the heat added may also be written as

$$H_3 - H_2 = (H_3 - H_S) + \lambda_S + (H_S' - H_2),$$

where $H_S = H_S' + \lambda_S$. Therefore,

$$H_3 - H_2 = C_p(T_3 - T_S) + \lambda_S + (C_p)_L(T_S - T_2),$$

where $(C_p)_L$ is the heat capacity of the liquid phase.
Since

$$H_3 - H_2 = C_p(T_3 - T_4) + (H_4 - H_R) + \lambda_R + (H_R' - H_1) - (-\Delta W_{\text{pump}})$$

and if $\lambda_S = \lambda_R$, $H_4 - H_R = 0$, and $H_R' - H_1 = 0$, then

$$C_p(T_3 - T_S) + (C_p)_L(T_S - T_2) = C_p(T_3 - T_4) - (-\Delta W_{\text{pump}})$$

or

$$(C_p)_L (T_S - T_2) = C_p(T_S - T_4) - (-\Delta W_p).$$

Neglecting $(-\Delta W_p)$, it would follow that if $T_S > T_2$, then $T_S > T_4$. Otherwise, the pumping requirement would be given by

$$H_2 - H_1 = (-\Delta W_p) = C_p(T_S - T_4) - (C_p)_L (T_S - T_2).$$

Since it can be expected that $C_p < (C_p)_L$, the preceding expression is indicative that $(T_S - T_4) > (T_S - T_2)$, or $T_2 > T_4$.

However, by the assumptions made, $T_4 = T_R = T_1$. In consequence, therefore, it is necessary that $T_2 > T_1$. In other words, the pumping requirement will be manifested by a slight increase in temperature of the liquid phase.

Efficiency. Interestingly, the ideal Rankine efficiency also may be stated as

$$\text{Eff}_{\text{Rankine}} = \frac{\text{work done}}{\text{heat added}} = \frac{\text{heat added} - \text{heat lost}}{\text{heat added}} = 1 - \frac{\text{heat lost}}{\text{heat added}}.$$

Substituting in terms of the stream enthalpies,

$$\text{Eff}_{\text{Rankine}} = \frac{(H_3 - H_2) - (H_4 - H_1)}{(H_3 - H_2)}$$

$$= \frac{(H_3 - H_4) - (H_2 - H_1)}{(H_3 - H_2)}$$

$$= \frac{(H_3 - H_4) - (-\Delta W_p)}{(H_3 - H_2)}.$$

The result is the same as before, if $(-\Delta W_p)$ includes $H_2 - H_1 = (-\Delta W_{\text{pump}})$.

Adjusting the Rankine Efficiency

The Rankine efficiency can be presented with the pumping losses appearing in both the numerator and denominator. Offsetting, the efficiency is lowered by the one, raised by the other, as if pumping losses tend to be recouped in the circulating working fluid. Using the previous notation, the value is

$$H_2 - H_1 = (-\Delta W_{\text{pump}}) \leq (-\Delta W_p)$$

It also can be argued that, if this is the case, then the pumping require-ments should simply be ignored in calculating the efficiency, since this is an internal matter: It all depends upon the convention chosen.

There is more to consider, however. The usual representation for the heat and work balance, in the differential form, is

$$dq - dw = dH = C_p \, dT - C_p \mu \, dP = V \, dP + d\text{lw},$$

where it is understood that the quantities are on a unit mass or unit mole basis. For a liquid, most usually, $\mu \to 0$ (albeit slightly negative). Hence the enthalpy change is a function of temperature only. If the pumping step operates essentially isothermally, then there is no enthalpic change. Ac-cordingly, $-dq = -dw$. The pumping work done on the system $(-\Delta W_{\text{pump}})$ would be dissipated as heat $(-\Delta Q)$. Moreover, the negative sign indicates that this would be heat lost to the surroundings. Thus $H_2 - H_1 = 0$.

The usual convention, therefore, is only to subtract parasitic require-ments from the theoretical work done, writing the Rankine efficiency simply as

$$\text{Eff}_{\text{Rankine}} = \frac{C_p(T_3 - T_4) - (-\Delta W_{\text{parasitic}})}{C_p(T_3 - T_4) + \lambda_R},$$

where $(-\Delta W_{\text{parasitic}})$ includes the pumping requirements and other desig-nated parasitic energy requirements.

Noncondensing Turbines

The cycle can be operated on an open or once-through basis using so-called noncondensing or extraction (back-pressure) turbines, e.g., for the steam–water system. The exhaust steam also will constitute a product, as process steam or for a second Rankine cycle operating at different conditions, i.e., at lower levels. The overall operation is commonly referred to as cogeneration.

The efficiency per se can again be represented by the same formula, or else the waste-heat loss for cooling (leg 4-1) can be discounted, since this waste-heat is otherwise utilized.

Binary Systems

The working fluid per se can be a binary or a multicomponent system also. The saturation curve would then be replaced by a two-phase envelope, and the initial and final temperatures for vaporization and condensation would be different.

Alternatively, the working fluid exhaust may be (partially) absorbed in a suitable liquid medium, pumped to pressure, and heated and revaporized to initiate the cycle again. The remaining unvaporized liquid is cooled externally to complete the cycle.

All of the foregoing factors, including the working fluid species, will affect the efficiency of the cycle. The species will at the same time affect working pressure–temperature levels.

The effects can be precalculated to a degree, with the final evaluation resting on experiment.

Fuel Efficiency

The first and foremost enabling feature of the Rankine cycle is the near-ambient temperature reached by the cold working fluid (T_1 or T_2). This means that the working fluid can be heated (and vaporized) counter-currently by the fuel combustion gases, whose exit or stack temperature may reasonably approach ambient conditions (say a 200–250°F approach), which will ensure satisfactorily high fuel efficiencies, on the order of 90%.

2.5. THE JOULE CYCLE

The Joule cycle per se will be discussed in detail in the next chapter. It involves the compression of a working fluid to a higher temperature, the addition of heat, the expansion to a lower temperature, and the rejection of heat. For the purposes of analysis, the compression and expansion steps may be regarded as isentropic. The addition and rejection of heat result in temperature changes in the working single-phase gaseous fluid, and are taken into account in the analysis of the cycle efficiency. Flow and P-T diagrams are provided in Fig. 2.6.

The Joule cycle normally operates as a closed cycle whereby the working fluid recirculates. If the Joule cycle is operated as an open cycle, the working fluid flows through on a once-through basis and the designator employed is the Brayton cycle, although the designators are sometimes interchanged. The notable embodiment of the latter is the gas-fired turbine.

Suffice to say, here, the theoretical efficiency of the Joule cycle is

$$\text{Eff}_{\text{Joule}} = 1 - \frac{1}{\alpha},$$

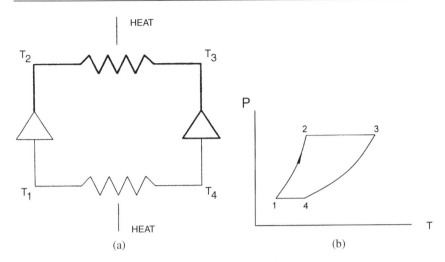

Fig. 2.6. Joule cycle. (a) Flow diagram; (b) *P-T* coordinates.

where α is the ratio of the absolute temperatures at the end and beginning of compression and is always greater than unity. The value of α is also necessarily the ratio of the absolute temperatures at the beginning and end of expansion, by virtue of operating between the same two pressure levels.

The theoretical efficiency of the Joule cycle may be a considerable way from the actual overall operating efficiency, due not only to compressor and expander losses, but due particularly to fuel efficiency losses, as will be discussed subsequently. Nevertheless, there is a way around this latter difficulty, as will be developed fully in Chapter 4: It involves the utilization of waste-heat from compression to heat the fuel combustants.

2.6. THE USE AND MISUSE OF ENTROPY

It is the custom at times to use the idea of entropy in the analysis of energy cycles, e.g., the Carnot, Rankine, and Joule cycles. It is a concept that is at the least superfluous and at the most may be misleading. Because entropy is a secondary variable or function, calculated from the primary variables of temperature and pressure or volume, it is thus a step removed from being a dynamic observable variable that describes system behavior —and its use may become more an exercise in obscurantism. Accordingly, aspects of its derivation and usage will be examined and critiqued.

Open or Flow Systems

In terms of the enthalpy function for a flow system, in consistent units, and as previously developed,

$$dq - dw = dH = \frac{\partial H}{\partial T} dT + \frac{\partial H}{\partial P} dP$$

$$= C_p \, dT - C_p \, \mu \, dP$$

$$= V \, dP + T \, dS,$$

where $T \, dS = d\omega$—the intrinsic energy change. For the purposes here, the equations are on a unit molar basis. Where required, it is understood that the mechanical equivalent of heat J_0 is implicitly introduced in order to make the units and terminology consistent.

For a perfect gas, the Joule–Thomson coefficient $\mu = 0$ and it follows that

$$C_p \, dT = V \, dP + T \, dS.$$

Furthermore, since for a perfect gas $V = RT/P$,

$$C_p d \ln T - R d \ln P = dS.$$

This is a principal reason for introducing dS or S, since the preceding equation will constitute a perfect differential and S will become a state function of T and P; that is, a separation of variables is achieved, and the integration will be independent of path, in other words, is independent of the behavior of T versus P. In short, $1/T$ serves as the integrating factor for a total differential equation in three variables.

The other reason, of course, is that S will be constant for an isentropic compression or expansion.

Integrating, where for convenience C_p is a constant (though it may also be a function of T but not of P),

$$S = C_p \ln T - R \ln P + \text{constant},$$

or, on integrating between limits,

$$S - S_0 = C_p \ln \frac{T}{T_0} - R \ln \frac{P}{P_0},$$

where here the subscript 0 denotes some arbitrary reference point or condition.

Alternately,

$$S - S_0 = \ln\left[\left(\frac{T}{T_0}\right)^{C_p}\left(\frac{P}{P_0}\right)^{-R}\right].$$

In terms of temperature,

$$\frac{T}{T_0} = \left(\frac{P}{P_0}\right)^{R/C_p} \exp\left(\frac{S - S_0}{C_p}\right).$$

The preceding equation would provide a plot of, say, T versus S for parameters of P. There would be a series of curves in T-S space for parameters of P.

For a nonideal gas, the integration may be denoted as

$$S = \int \frac{1}{T}[dH - V\,dP] + \text{constant}.$$

In general H will be a function of both T and P, and V can be represented as $V = zRT/P$ where the compressibility factor z is a function of both T and P. Integration will require a knowledge of the behavior of both H and z, and numerical procedures will no doubt be required. The calculation is greatly simplified if P is held constant or if T is held constant. The calculations for steam, for instance, have adopted these procedures, as explained in the introductory chapter to the steam tables [(1), p. 23]. The difficulty lies in the fact that these determinations are incomplete, since, in general, entropy is not a state function of temperature and pressure.

Thus the determinations in the superheated vapor region are at constant pressure and temperature [(1), Table 3] and the determinations for compressed water are also at constant pressure and temperature [(1), Table 4]. It must be borne in mind, however, that these are specific paths which are so ordained and which are in a strict sense limited to either constant P or constant T. Other specified paths of behavior between T and P would yield other patterns of behavior for the entropy S, each pattern specific to the assumed path of behavior between T and P. In other words, rigorously speaking, entropy depends upon the path of variation between T and P, and properly is not a state function independent of path. The most significant path of behavior between T and P is along the vapor-pressure or saturation curve, which will be discussed subsequently.

Closed Systems

As developed previously,

$$dq - dw = dU = \frac{\partial U}{\partial T} dT + \frac{\partial U}{\partial V} dV$$

$$= C_v \, dT - C_v \eta \, dP$$

$$= -P \, dV + T \, dS,$$

where $T \, dS = d\omega$—the intrinsic energy change. Where required, it is again understood that the mechanical equivalent of heat J_0 is implicitly introduced in order to make the units and terminology consistent.

For a perfect gas, the Gay–Lussac coefficient $\eta = 0$ and it follows that

$$C_v \, dT = -P \, dV + T \, dS.$$

Furthermore, since for a perfect gas $P = RT/V$,

$$C_v d \ln T + Rd \ln V = dS.$$

This again is a principal reason for introducing dS or S, since the preceding equation will constitute a perfect differential and S will become a state function of T and V. A separation of variables is achieved, and the integration will be independent of path, that is, independent of the behavior of T versus V. Furthermore, S will be constant for an isentropic compression or expansion.

For the record, integration may proceed as before. Thus assuming for convenience that C_v is a constant (though it may be a function of T but not of V),

$$S = C_v \ln T + R \ln V + \text{constant}$$

or

$$S - S_0 = C_v \ln \frac{T}{T_0} + R \ln \frac{V}{V_0},$$

where again the subscript 0 denotes some arbitrary reference point or condition.

Alternately,

$$S - S_0 = \ln\left[\left(\frac{T}{T_0} \right)^{C_v} \left(\frac{V}{V_0} \right)^{R} \right]$$

and in terms of temperature,

$$\frac{T}{T_0} = \left(\frac{V}{V_0}\right)^{-R/C_v} \exp\left(\frac{S - S_0}{C_v}\right).$$

The preceding equations would provide a plot of say T versus S for parameters of V. Thus there would be a series of curves in T-S space for parameters of V.

For a nonideal gas, the integration may be denoted as

$$S = \int \frac{1}{T}[dU + P\,dV] + \text{constant.}$$

In general, U will be a function of both T and V, and P may be represented as $P = zRT/V$, where the compressibility factor z can be made a function of both T and V. Integration will require a knowledge of the behavior of both U and z, and numerical procedures will no doubt be necessary. The calculation is greatly simplified if V is held constant or if T is held constant.

The results for open and closed systems are entirely equivalent. Thus consider the relationships developed for a perfect gas:

$$C_p d \ln T - Rd \ln P = dS$$

and

$$C_v d \ln T + Rd \ln V = dS.$$

Combining to eliminate dS,

$$(C_p - C_v)d \ln T = R(d \ln P + d \ln V)$$

or

$$Rd \ln T = Rd \ln PV.$$

Thus

$$d \ln T = d \ln PV,$$

which integrates to yield

$$\ln T = \ln PV + \text{constant.}$$

If the constant of integration is chosen so that constant $= -\ln R$, then

$$RT = PV,$$

which is the original assumption. The relationships are fully consistent. Thus the same results would have to be achieved whether represented in terms of T and P or in terms of T and V. Another way of saying it is that if S is known as a function of T and P, then substitution of the ideal gas law will yield S as a function of T and V, and vice versa.

The analogy may also be extended to nonideal systems, although the independent behavior of T versus P or T versus V must be specified first. As limiting cases, T may be held constant or P or V may be held constant, depending upon whether the system is to be considered open or closed.

Thus, in general, for any system—ideal or nonideal—and succinctly put,

$$dH - V\,dP = T\,dS,$$
$$dU + P\,dV = T\,dS.$$

If the term $T\,dS$ is equal in the preceding two equations, then

$$dH = dU + d(PV),$$

which is the prescribed relationship between dH and dU. The consistency is already built in, so to speak.

Carnot Cycle

The Carnot cycle is readily diagrammed in T-S space as a rectangle, as shown schematically in Fig. 2.7. The vertical lines represent isentropic compression and expansion, whereas the horizontal lines represent the isothermal addition of heat at T_h and the isothermal rejection of heat at T_c. If the ordinate T starts at absolute zero, then the temperatures are denoted by the lengths of the respective vertical lines at constant entropy, whereby

$$\text{Eff}_{\text{Carnot}} = \frac{T_h - T_c}{T_h}.$$

The areas of the enclosing rectangles could instead be used to represent the efficiency, since the respective areas are directly proportional to the vertical distance for each triangle. Note also that the horizontal distances are related to the compression and expansion ratios, as per the derivation for the Carnot efficiency.

It should be stressed, however, that areas of the enclosing rectangles do not represent energy change as such; that is, if an area is represented by the integral $\int T\,dS$ or by the difference in the integrals of two areas, this

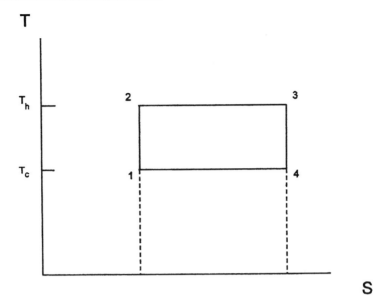

Fig. 2.7. Schematic T-S diagram for the Carnot cycle.

area will have the dimensions of energy but will not constitute enthalpic or internal energy per se, since

$$T\,dS = dH - V\,dP \quad \text{or} \quad T\,dS = dU + P\,dV.$$

Only if P is constant in the first case will $\int T\,dS$ represent enthalpic energy change; in the second case, only if V is constant will $\int T\,dS$ represent internal energy change.

There is the temptation to extend the workings of the Carnot to systems undergoing a phase change, e.g., vaporization and condensation, as set forth by Hougen *et al.* [(2), p. 728]. Thus consider Fig. 2.8, where the phase envelope in T-S space is represented schematically by the bell-shaped curve. To the left of the envelope is the supercooled liquid region; to the right is the superheated vapor region. The interior of the envelope is the two-phase vapor–liquid region. The apex of the envelope constitutes the critical point.

The left-hand curve would represent the entropy of the saturated liquid phase; the right-hand curve would be the entropy of the saturated vapor phase. At a constant temperature, the difference between the curves is given by $\Delta S = \lambda/T$, where λ is the latent heat of vaporization.

If a Carnot cycle could be inserted within the envelope as also shown in Fig. 2.8, then let leg 1-2 denote an isentropic change, followed by an

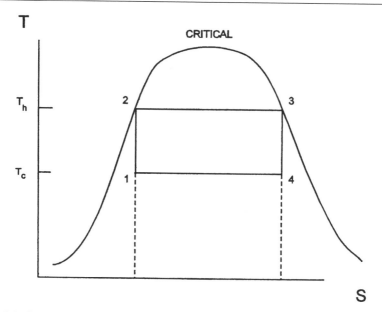

Fig. 2.8. Schematic representation of a Carnot cycle inside the two-phase envelope in *T-S* coordinates.

isothermal vaporization at T_h along leg 2-3. In turn, there is an isentropic change along leg 2-3, followed by an isothermal condensation at T_c along leg 4-1.

There are, however, exceptions which must be critiqued. Most importantly, the Carnot cycle pertains to changes which occur in a single phase or in the single-phase region. Moreover, the single phase is what is ordinarily thought of as a gas or superheated vapor. This fact is implicit in the derivations for the Carnot efficiency.

In the foregoing cycle both vaporization and condensation occur at varying liquid–vapor ratios. More significantly, compression up through the two-phase region will, as expected, be at a varying liquid–vapor ratio. Similarly, expansion down through the two-phase region also will be a varying liquid–vapor ratio. (Not to mention the effect of condensate on turbine blades if that is to be the embodiment for the expansion.) There is the additional complication that the one phase may be dispersed within the other, or vice versa, or else the phases will exist separately—each a distinct continuum.

We further comment that the entropy of the saturated vapor follows the vapor saturation curve in *T-S* space and the entropy of the saturated liquid follows the liquid saturation curve in *T-S* space. Although the liquid and vapor entropies may be prorated at a common temperature based on the liquid or vapor content, and can so be made to give a prorated or averaged value, it does not follow that the compression or expansion of the liquid–vapor mixture will be isentropic as a whole or if it even has meaning in such terms. Additionally, latent heat effects will be involved because condensation occurs during compression and vaporization occurs during expansion. All in all, the picture is very complicated, and such simplifications as assuming a path of constant entropy may be far afield.

Moreover, although vaporization and condensation occur at constant temperature, the pressure also will be constant—a constraint which does not occur in the ordinary presentation of the Carnot cycle and the calculation of its efficiency.

In summary, while we may speak of the isentropic compression and expansion of a gas or superheated vapor, what do we mean if a more-dense, liquid phase is present? Alternatively, if the phase is all liquid, what then? It therefore becomes necessary to extend our concept of a cycle to what is referred to as the Rankine cycle, which accommodates these changes in phase. The embodiment of the Rankine cycle, perforce, is an open or flow system.

Rankine Cycle

Hougen *et al.*, for example, have made a comparison of the Carnot and Rankine cycles in *T-S* space utilizing the two-phase envelope, without superheating the vapor, in which they purport to demonstrate graphically that the Carnot cycle will always exhibit a higher efficiency than the Rankine cycle [(2), pp. 742–743]. The evidence is questionable, however, for the reasons already mentioned, namely, that the Carnot cycle pertains only to a gaseous or vapor-phase system rather than to a two-phase system. In subsequent pages, Hougen *et al.* show the effect of superheating the vapor on a *T-S* diagram [(2), pp. 745–749].

The Rankine cycle may be represented more exactly and completely in *T-S* space as shown schematically in Fig. 2.9. Leg 1-2 signifies the pumping of liquid condensate from a lower pressure P_A to a higher pressure P_B. Leg 2-S' is the heating of the condensate at varying temperature to the vaporization temperature $T'_S = T_S$. This step involves sensible heat changes in the liquid phase. Leg S'-S denotes the vaporization at constant temperature. Leg S-3 is the superheating of the vapor, at varying temperature, which involves sensible heat changes in the vapor phase. This superheating is followed by an isentropic expansion from pressure P_B down to pressure P_A, which constitutes leg 3-4.

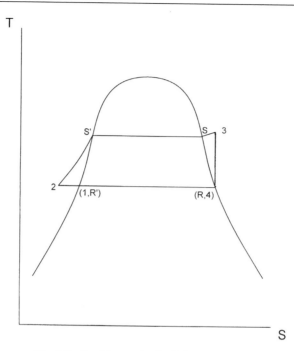

Fig. 2.9. Rankine cycles in *T-S* coordinates.

Note that the isentropic expansion that is represented by the vertical line at constant entropy, here intersects the saturated vapor curve at $T_4 = T_R = T_R'$. It could as well be represented such that $T_4 > T_R = T_R'$, and a further cooling of the vapor would be required between T_4 and $T_R = R_R'$. The important fact is that for the isentropic expansion to intersect the saturated vapor curve or stay to the right of the curve, the vapor at T_3 must be *superheated*.

Condensation at a constant temperature is denoted by leg *R-R'*. While supercooling of the condensate may be incurred, for the purposes here no supercooling is shown.

If pumping is regarded as essentially an isothermal process and the Joule–Thomson coefficient for a liquid is negligible, then

$$T\,dS = -V\,dP \quad \text{and} \quad \Delta S = -\frac{1}{T}V(P_B - P_A),$$

where V is the specific volume, which is essentially a constant. Thus ΔS is negative and is so represented isothermally on the diagram that is Fig. 2.9.

For sensible heat changes in a single phase, liquid or gaseous,

$$T\,dS = C_p\,dT - C_p\,\mu\,dP - V\,dP.$$

If the pressure is constant, then

$$dS = C_p\frac{dT}{T} \quad \text{and} \quad \Delta S = \int C_p\frac{dT}{T} + \text{constant},$$

and if C_p is constant, then integrating between limits,

$$\Delta S = S - S_0 = C_p \ln\left(\frac{T}{T_{\text{ref}}}\right) \quad \text{or} \quad \frac{T}{T_0} = \exp\left(\frac{S - S_0}{C_p}\right),$$

where here $T - T_{\text{ref}}$ is the difference in the temperature levels. Thus in T-S space the behavior of T in terms of S would be exponential, with T increasing with S. This trend is represented schematically in Fig. 2.9 by the curved line 2-S' (leg 2-S') for heating the pressurized condensate and by curved line S-3 (leg S-3) for superheating the vapor produced.

This all said, the efficiency of the Rankine cycle still must be determined by the methods previously presented, in terms of enthalpic behavior. For purposes of comparison, a schematic representation of the Rankine cycle is shown in T-H coordinates in Fig. 2.10 and in its more usual display using P-H coordinates in Fig. 2.11.

Joule Cycle

The Joule cycle can be conveniently represented on a P-S diagram as shown in Fig. 2.12.

Heat is added at a constant pressure P_B along leg 2-3 and in part removed at a constant pressure P_A along leg 4-1. Isentropic compression occurs along leg 1-2 and isentropic expansion occurs along leg 3-4.

Observe that at constant pressure,

$$dH = C_p\,dT = T\,dS.$$

If, say, C_p is constant, then

$$S_3 - S_2 = C_p \ln\frac{T_3}{T_2}$$

$$S_4 - S_1 = C_p \ln\frac{T_4}{T_1},$$

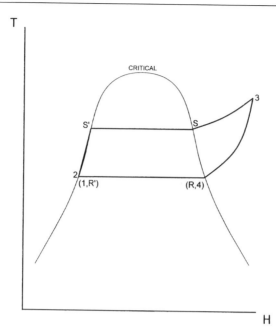

Fig. 2.10. *T-H* diagram for the Rankine cycle.

and

$$\ln\frac{T_3}{T_2} = \ln\frac{T_4}{T_1}$$

or

$$\frac{T_3}{T_2} = \frac{T_4}{T_1} \quad \text{and} \quad \frac{T_4}{T_3} = \frac{T_1}{T_2} = \frac{1}{\alpha}.$$

Therefore,

$$1 - \frac{T_4}{T_3} = 1 - \frac{T_1}{T_2} = 1 - \frac{1}{\alpha}$$

and

$$\text{Eff}_{\text{Joule}} = \frac{T_3 - T_4}{T_3} = \frac{T_2 - T_1}{T_2} = 1 - \frac{1}{\alpha}.$$

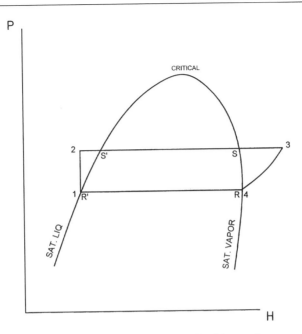

Fig. 2.11. *P-H* diagram for the Rankine cycle.

Thus the efficiency is based on the temperature levels between isobaric and isentropic behavior and is of the same form as the Carnot efficiency.

Joule Cycle for a Closed System. Alternately, the Joule cycle can be viewed as pertaining to a closed system, described in terms of the variables T and V. In this case, the cycle will appear as a rectangle in V-S coordinates.

Corresponding to the flow or open system representation and starting again at a point 1, there is an isentropic compression from a larger volume V_A to a smaller volume V_B at point 2 along leg 1-2. Heat is added at the constant volume V_B along leg 2-3. This is followed by an isentropic expansion along leg 3-4, from volume V_B to volume V_A. Heat is rejected at the constant volume V_A along leg 4-1 to complete the cycle.

At constant volume,

$$dU = C_v \, dT = T \, dS.$$

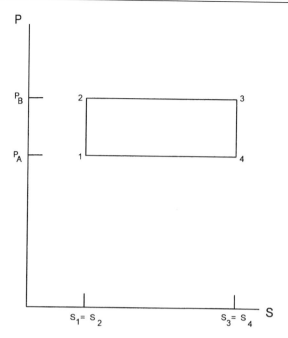

Fig. 2.12. Schematic *P-S* diagram for the Joule cycle.

If, say, C_v is constant, then

$$S_3 - S_2 = C_v \ln \frac{T_3}{T_2},$$

$$S_4 - S_1 = C_v \ln \frac{T_4}{T_1},$$

and

$$\frac{T_3}{T_2} = \frac{T_4}{T_1}.$$

Exactly the same results are reached as for an open system.

We add in passing that the Carnot cycle can be represented as well for an open system as for a closed system.

Construction of a *T-S* Diagram

The locus of reference for the construction of a *T-S* diagram is the vapor-pressure curve, or saturation curve. Here, there is a relationship between T and P, both intensive variables, that makes it possible to present the T-S diagram whether the system is ideal or nonideal.

For an ideal gas, the results have been derived already as

$$S - S_0 = \ln\left[\left(\frac{T}{T_0}\right)^{C_p}\left(\frac{P}{P_0}\right)^{-R}\right]$$

or

$$\frac{T}{T_0} = \left(\frac{P}{P_0}\right)^{R/C_p}\exp\left(\frac{S - S_0}{C_p}\right).$$

The vapor-pressure curve relates temperature and pressure, that is, say, in the terms of pressure as a function of temperature, which is symbolically expressible as $P = P(T)$. The substitution of this relationship into the foregoing expression for the entropy difference would yield the entropy of the saturated vapor as a function of temperature only or, inversely, temperature as a function of the saturated vapor entropy only.

For a nonideal gas, the integration has been expressed already as

$$S = \int \frac{1}{T}[dH - V\,dP] + \text{constant},$$

where, in general, H will be a function of both T and P, and $V = zRT/P$, where the compressibility factor z is a function of both T and P. If the vapor pressure curve behavior is known as $P = P(T)$, then substitution will permit the numerical integration of the preceding expression to yield the T-S diagram for the saturated vapor.

(Other representations for the gas law or equation of state can be used; some are more convenient than others. Suffice it to say that there is a relationship between P, T, and V which in some formulations can be rather complex to use; that is, V is not readily obtained as an explicit function only of T and P.)

The entropic behavior of the saturated liquid is obtained by difference; that is, at each temperature, the entropy difference for vaporization is given by $\Delta S = \lambda/T$. This value may be subtracted from the corresponding value for the saturated vapor entropy. Since enthalpy is regarded as an additive property and may be prorated for the liquid or vapor content, then entropy may be similarly prorated in the two-phase region at a constant temperature (and pressure) inasmuch as $\Delta S = \Delta H/T$.

In this manner, the T-S diagram for a saturated vapor–liquid system can be built up. The determination will require designation of the baseline values for the variables and functions which have been subscripted with 0. The steam tables use the following baseline values [(1), Table 1]:

$T_0 = 32°F$ (or 492°R),

$P_0 = 0.08854$ psia (the vapor pressure at 32°F),

H_0 (saturated liquid) = 0.00 Btu/lb (or 0.00 Btu/lb-mol),

H_0 (saturated vapor) = 1075.8 Btu/lb (or 19,364.4 Btu/lb-mol),

S_0 (saturated liquid) = 0.0000 Btu/lb-°R (or 0.0000 Btu/lb-mol-°R),

S_0 (saturated vapor) = 2.1877 Btu/lb-°R (or 39.3886 Btu/lb-mol-°R).

The determination of the entropic behavior in the single-phase regions is something else again. As mentioned previously, the steam tables use constant temperature in the supercooled liquid region for the entropic determination and use constant pressure in the superheated vapor region. Both assumptions are conveniences that avoid the unfortunate fact that there are three variables (S, T, P) and only one equation. Only if an ideal gas is assumed in the superheated vapor region can this unfortunate fact be circumvented, by permitting a separation of variables as already shown.

Similar sorts of problems emerge in the supercooled liquid region.

Entropy and Nonideality

The discussion will be confined to open or flow systems, in terms of the enthalpy H. The changes at both constant pressure and constant temperature will be examined, whereby $(\partial H/\partial T)_P = C_p$ and $(\partial H/\partial P)_T = -C_p \mu$.

The relationship between enthalpy and entropy at constant P may be expressed as

$$dH|_P = C_p|_P \, dT = T \, dS.$$

Rearranging and integrating at constant P,

$$\int_{T_{ref}}^{T} C_p|_P \frac{dT}{T} = (S - S_{ref})|_P.$$

Therefore, at constant pressure, S is a function of T or, inversely, T is a function of S. Symbolically, $S = S(T)|_P$ or, inversely, $T = T(S)|_P$. Thus we can speak of enthalpy being a function of entropy whereby, at constant pressure,

$$(H - H_{ref})|_P = \int_{S_{ref}}^{S} T(S)|_P \, dS.$$

Therefore, the enthalpic change at constant pressure may be related to the entropic behavior at constant pressure. For the purposes here, the subscript "ref" will pertain to a point on the saturated vapor curve in T-S coordinates, whereby the specification of a constant pressure P_{ref} affixes a value for T_{ref} from the vapor-pressure curve, and which in turn affixes S_{ref}.

At constant temperature,

$$dH|_T = -C_p\mu|_T\, dP = V|_T + T|_T\, dS,$$

where the constant temperature $T|_T$ may be assigned a value $T|_T = T_{\text{ref}}$, which has a corresponding value P_{ref} on the vapor-pressure curve and will in turn have a corresponding value of T_{ref} on the saturated vapor curve in T-S coordinates. Rearranging and integrating,

$$\int_{P_{\text{ref}}}^{P} \left[-C_p\mu|_T - V|_T \right] dP = T_{\text{ref}}(S - S_{\text{ref}})|_T.$$

Therefore, at constant temperature, S becomes a function of P or, inversely, P becomes a function of S. Symbolically, $S = S(P)|_T$, or, inversely, $P = P(S)|_T$. It follows that

$$(H - H_{\text{ref}})|_T = \int_{P(S_{\text{ref}})|_T}^{P(S)|_T} V|_T\, dP + T_T(S - S_{\text{ref}})|_T.$$

Thus the enthalpic change at constant temperature may be related to the entropic behavior at constant temperature.

The foregoing derivations also can be made for the supercooled liquid region using the saturated liquid curve from the T-S diagram for the reference condition, both at constant pressure and constant temperature. For a liquid, it is an appropriate assumption that $\mu \to 0$ and that the specific volume $V = 1/\rho$ is a constant. Therefore,

$$dH = C_p\, dT = \frac{1}{\rho}\, dP + T\, dS.$$

At constant pressure, integrating from, say, $T_{\text{ref}} = T_0$,

$$dH|_P = C_p|_P\, dT = T\, dS$$

and

$$\int_{T_{\text{ref}}}^{T} C_p|_P \frac{dT}{T} = (S - S_{\text{ref}})|_P.$$

If C_p is constant,

$$(S - S_{\text{ref}})|_P = C_p \ln \frac{T}{T_0} \quad \text{or} \quad \frac{T}{T_0} = \exp\left(\frac{S - S_0}{C_p}\right).$$

Therefore, at constant pressure, S is again a function of T or, inversely, T is a function of S. Symbolically, $S = S(T)|_P$ or, inversely, $T = T(S)|_P$. Thus we can again speak of enthalpy being a function of entropy whereby at constant pressure,

$$(H - H_{\text{ref}})|_P = \int_{S_{\text{ref}}}^{S} T(S)|_P \, dS.$$

At constant temperature,

$$dH|_T = 0 = V|_T \, dP + T|_T \, dS$$

and

$$(S - S_{\text{ref}})|_T = \frac{1}{T}|_T\left[-\frac{1}{\rho}(P - P_{\text{ref}})\right]\bigg|_T$$

$$\text{or} \quad (P - P_{\text{ref}})|_T = -T\rho(S - S_{\text{ref}})|_T.$$

The entropic behavior of a supercooled liquid at constant temperature will therefore be independent of the enthalpy.

The problem with the foregoing several considerations is that the determinations for entropy or entropy difference will not be independent of path, that is, in this case independent of whether constant pressure is assumed or constant temperature is assumed. Thus consider Fig. 2.13 as pertains to the saturated vapor curve in T-S space. Let the determination at constant temperature start at the reference condition or point (T_a, P_a, S_a) and end at the point (T_a, P_b, S). Let the determination at constant pressure start at the point (T_b, P_b, S_b) and end at (T_a, P_b, S). The terminal value S is required to be the same for both routes.

At constant temperature, where $T = T_a$, and in the appropriate and corresponding symbols,

$$\frac{1}{T_a} \int_{P_a}^{P_b} \left[-C_p \mu|_T - V|_T\right] dP = (S - S_a)|_T.$$

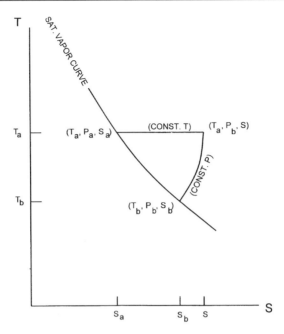

Fig. 2.13. Schematic comparison of entropy determinations at constant temperature and constant pressure.

At constant pressure, where $P = P_b$, and in the appropriate and corresponding symbols,

$$\int_{T_b}^{T_a} C_p|_P \frac{dT}{T} = (S - S_b)|_P.$$

There is, in general, simply no way that the value of S can be the same in both of the preceding expressions, since S_a and S_b are arbitrary, and the behavior patterns of C_p, μ, and V are arbitrary. For example, any of several equations of state can serve to express the volume or specific volume as a function of temperature and pressure.

Another way of looking at the problem is to regard the situation in P-T space as shown in Fig. 2.14 for the Rankine cycle. Here, the entropy determinations at constant pressure can be considered relative to the two-phase vapor-pressure curve. As examples, the entropy behavior could be determined at a constant pressure P_A starting at the saturation curve

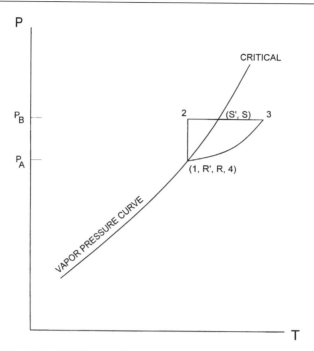

Fig. 2.14. *P-T* diagram for the Rankine cycle where $T_1 = T_4$.

and at a constant pressure P_B starting at the saturation curve. Then, say, consider an arbitrary entropy value at pressure P_A and temperature T_A designated as S_A, and consider another entropy value at pressure P_B and temperature T_B designated as S_B. The object is to examine the effect of path in calculating from S_A to S_B; that is, is $S_B - S_A$ independent of path?

Thus, for the two alternative routes, partial integration may be performed first at constant pressure and then at constant temperature, or vice versa, as follows:

$$S_B - S_A = \int_{T_A}^{T_B} [C_p/T]|_{P_A}\, dT + \int_{P_A}^{P_B} \left[-\frac{C_p\,\mu}{T} - \frac{V}{T} \right] \bigg|_{T_B}\, dP,$$

$$S_B - S_A = \int_{P_A}^{P_B} \left[-\frac{C_p\,\mu}{T} - \frac{V}{T} \right] \bigg|_{T_A}\, dP + \int_{T_A}^{T_B} \left[\frac{C_p}{T} \right] \bigg|_{P_B}\, dT.$$

The two different routes for integration cannot produce the same results unless the function dS is a perfect or exact differential.

Perfect and Imperfect Differentials

The problem in essence is that, in general, dS is not a perfect or exact differential. The differential may be stated as

$$dS = \left[\frac{C_p}{T}\right] dT + \left[-\frac{C_p \mu}{T} - \frac{V}{T}\right] dP.$$

The requirement for a perfect differential is that the order of partial differentiation be immaterial, whereby

$$\frac{\partial^2 S}{\partial P \partial T} = \frac{\partial^2 S}{\partial T \partial P}.$$

For single-phase behavior, this will not, in general, be the situation since

$$\frac{\partial[-C_p \mu/T - V/T]}{\partial P} \neq \frac{\partial[C_p/T]}{\partial T};$$

that is, the terms in each bracket behave independently of one another. Since C_p, μ, and V are, in general, all a priori functions of T and P, then if the equality existed, it would set up a correspondence between T and P, whereas T and P are independent variables. Only if the assumption of a perfect gas is made will

$$\frac{\partial^2 S}{\partial P \partial T} = 0 = \frac{\partial^2 S}{\partial T \partial P}.$$

Thus entropic determinations made at either constant pressure or constant temperature cannot, in general, produce or establish entropy as a point or state function of temperature and pressure, that is, symbolically, $S \neq S(T, P)$. Only if the assumption of a perfect gas is made can S be so represented, symbolically as $S = S(T, P)$. Likewise for a supercooled liquid. If $\mu = 0$ and $V = 1/\rho$ is a constant, then the inequality will again be sustained; that is,

$$\frac{\partial^2 S}{\partial P \partial T} = \frac{\partial[C_p/T]}{\partial P} \neq \frac{\partial^2 S}{\partial T \partial P} = 0.$$

The foregoing doubt casts suspicions about entropic information presented for a real or nonideal gas. If the gas does not depart greatly from ideality, however, the determinations will probably suffice. Alternately viewed, why not simply assume an ideal or perfect gas at the very start?

By contrast, the nature of enthalpy is that of a state function. As previously derived, if the Joule–Thomson coefficient μ is known as a function of temperature and pressure, then symbolically,

$$\mu = \frac{dT}{dP} = \mu(T, P),$$

which is a differential equation in terms of T and P. Rearranging and introducing an integrating factor C_p,

$$0 = C_p\, dT - C_p\, \mu\, dP,$$

the expression, in principle, integrates analytically to the symbolic expression

$$H = H(T, P),$$

where H is introduced as a constant of integration and may be generalized as a function; that is, specifying a value for H will reproduce a particular isenthalp. In the foregoing sense, the symbol H denotes absolute enthalpy.

Furthermore, for a perfect or exact differential there is the requirement that the order of partial differentiation be irrelevant; that is,

$$\frac{\partial^2 H}{\partial P\, \partial T} = \frac{\partial^2 H}{\partial T\, \partial P}.$$

Substituting,

$$\frac{\partial C_p}{\partial P} = \frac{\partial(-C_p\, \mu)}{\partial T}.$$

In other words, the integrating factor C_p is to be a function of T and P such that the preceding equivalence is sustained.

The enthalpy function is ordinarily determined as a difference, however, based on experimental data for C_p and μ—a fact which requires numerical integration procedures. Thus,

$$H - H_0 = \int_{T_0}^{T} C_p|_{P_0}\, dT - \int_{P_0}^{P} C_p\, \mu|_T\, dP.$$

As indicated, the numerical integration may be performed first at a constant pressure P_0 and then at a constant pressure T. Alternatively, the numerical integration may be performed first at a constant temperature T_0 and then at a constant pressure P. Either way, the results are theoretically the same and will be the same provided the data are consistent, which is always more or less a problem, especially if calculated values for μ are used instead of experimental values. Note that, in general, both C_p and μ are functions of both temperature and pressure.

In consequence, the determination at constant pressure is not in itself sufficient to establish the complete behavior of H, nor is the determination at constant temperature alone sufficient to establish the complete behavior of H. Both determinations are necessary—the one following the other.

Another way of looking at it is that the preceding text explains why the determinations for entropic behavior are incomplete in the single-phase region. If entropy were a state function, then the successive partial integrations at both constant pressure and constant temperature would be required to fully establish the behavior of the entropy as a function of temperature and pressure. Obviously this is not the case.

A further detailing and explanation of the situation for single-phase entropic behavior is as follows. Since

$$dS = \frac{1}{T} dH - \frac{V}{T} dP$$

$$= \frac{1}{T} C_p \, dT + \frac{1}{T} [-C_p \mu - V] \, dP,$$

then if S is to be a state function, it would be required that

$$\frac{\partial S}{\partial T} = \frac{1}{T} C_p,$$

$$\frac{\partial S}{\partial P} = \frac{1}{T} [-C_p \mu - V],$$

and

$$\frac{\partial^2 S}{\partial P \, \partial T} = \frac{\partial^2 S}{\partial T \, \partial P}.$$

Substituting and performing the operations,

$$\frac{1}{T} \frac{\partial C_p}{\partial P} = \frac{1}{T} \frac{\partial(-C_p \mu)}{\partial T} - \frac{1}{T^2} C_p \mu + \frac{V}{T^2} - \frac{1}{T} \frac{\partial V}{\partial T}.$$

Since by virtue of the enthalpic behavior,

$$\frac{1}{T}\frac{\partial C_p}{\partial P} = \frac{1}{T}\frac{\partial(-C_p\mu)}{\partial T},$$

it would be required that

$$-\frac{1}{T^2}C_p\mu + \frac{V}{T^2} - \frac{1}{T}\frac{\partial V}{\partial T} = 0$$

or

$$-\frac{1}{T}C_p\mu + \frac{V}{T} - \frac{\partial V}{\partial T} = 0.$$

This would constitute a relationship between T and P, whereas by definition they are independent variables. A contradiction is produced.

For an ideal gas, however, $\mu = 0$ and $V = RT/P$ such that the foregoing relationship is satisfied.

For a liquid, if $\mu \sim 0$ and the specific volume or density essentially remains constant, then the relationship is not satisfied, since $V/T \neq 0$.

The foregoing considerations bring up the possibility that, in general,

$$dH = P\,dV + T\,dS$$

is not an allowable representation for a single-phase nonideal system nor is

$$dH = P\,dV + d\omega$$

an allowable representation; that is, put another way, the relationships are, in general, nonintegrable.

Mollier Chart

A Mollier chart is fundamentally a plot of enthalpy and entropy for parameters of both constant pressure and constant temperature. Enthalpy is plotted as the ordinate and entropy as the abscissa. Given the temperature and pressure—that is, where a constant temperature parameter intersects a constant pressure parameter—the enthalpy is determined. In essence, the temperature and pressure would be coordinates in x-y space, that is, in the plane, whereas the enthalpy would be the coordinate in z space, such that z is a function of x and y or, symbolically, $z = z(x, y)$. The parametric representation of constant temperature and pressure is fitted to the data and adjusted so that the enthalpy coordinate is linear with distance. Such a representation is a marvel of geometry.

At the same time and given the same temperature and pressure, the entropy is supposedly determined. (The angle between the enthalpic and entropic coordinates may be adjusted to assist the representation. Again it is all a marvel of the geometer's art.) In other words, entropy is treated as if it were a state function of the independent variables temperature and pressure. The foregoing proposition would therefore presumably establish a correspondence with enthalpy and entropy. While this is correct enough along the two-phase vapor-pressure envelope, such is not, in general, the case in the single-phase region save for a perfect gas.

The two-phase saturated liquid–vapor region is included, with parameters for the liquid content. The temperatures and pressures which follow the saturation envelope correspond to the vapor-pressure curve. In the two-phase region, the enthalpies for the saturated vapor and saturated liquid at equilibrium may be prorated based on the liquid or vapor content, since enthalpies are additive. Inasmuch as at a constant temperature $\Delta S = \Delta H / T$, the two-phase entropies also may be so prorated based on the liquid or vapor content. The superheated vapor region also contains parameters for the degrees of superheat; the supercooled liquid region is of less interest.

We further comment that

$$dH = V\,dP + T\,dS$$

establishes a connection between enthalpy and entropy—the one apparently being some kind of a function of the other—with constant temperature and constant pressure serving as parameters. Given the behavior of S, say, the behavior of H will follow. Although dH is a perfect differential in terms of T and P, the entropy differential dS, in general, cannot be—save for a perfect gas. The problems lie in the relationships for establishing the behavior of S in terms of temperature and pressure, which are necessarily imperfect and incomplete and depend upon the route or path chosen for the determination. While in the two-phase saturated region T and P are related by the vapor-pressure curve, there is no such restriction or constraint in the single-phase region. Thus there is, in general, the inconsistency in the single-phase region that if dH is a perfect differential, then dS cannot be a perfect differential, since the foregoing equation would provide the contradiction. Only for a perfect gas can dH and dS both be perfect differentials, since, in this simplification, $dH = C_p\,dT$, where C_p is a constant or a function of T only and $\mu = 0$, whereby, as previously derived,

$$dS = C_p\,d\ln T - R\,d\ln P,$$

the latter being the consequence of solving the total differential equation formed from the substitutions and rearrangements for a perfect gas.

The determinations and results as presented in a Mollier chart or diagram are not above suspicion, therefore, for the reasons that have been discussed. This said, the plots and cross-plots which are required to make up a Mollier chart are a formidable undertaking. The most well known Mollier chart is the chart for steam which accompanies the steam tables of Keenan and Keyes (1). The same remarks for the Mollier diagram of course apply to the steam tables as well because the Mollier diagrams are based on the steam tables.

A few summary comments, therefore, are in order: First, it may be presumed that the plots for enthalpy as a function of temperature are substantially correct, at least as far as the heat capacity and Joule–Thomson data used are correct. Furthermore, the entropic behavior along the saturated two-phase envelope is correct, as is the proration within the two-phase region in terms of liquid content.

In the superheated single-phase region, the entropic behavior may be judged reasonably correct at least to the degree that the gaseous system approximates a perfect gas. As a further explanation, for a perfect gas, as per the preceding relationship and as developed before,

$$\frac{\partial S}{\partial T} = \frac{C_p}{T},$$

$$\frac{\partial S}{\partial P} = -\frac{R}{P},$$

and

$$\frac{\partial^2 S}{\partial P \, \partial T} = 0 = \frac{\partial^2 S}{\partial T \, \partial P}$$

so that the order of differentiation is immaterial. Therefore, with respect to any reference point "ref,"

$$S - S_{\text{ref}} = [S(T, P_{\text{ref}}) - S_{\text{ref}}]|_{P_{\text{ref}}} + [S(T, P) - S(T, P_{\text{ref}})]|_T$$

$$= \int_{T_{\text{ref}}}^{T} \frac{C_p}{T}|_{P_{\text{ref}}} \, dT - \int_{P_{\text{ref}}}^{P} \frac{R}{P}|_T \, dP$$

$$= C_p \ln\left(\frac{T}{T_{\text{ref}}}\right)|_{P_{\text{ref}}} - R \ln\left(\frac{P}{P_{\text{ref}}}\right)|_T$$

and the order of integration is immaterial.

In the supercooled liquid region, the entropic determinations also are not exact; that is, the entropy cannot be a state function since the system is nonideal.

2.7. OTHER POWER CYCLES

The expansion step per se does not denote a power cycle, but must be coupled with a compression step (as in the Joule cycle) or with a pumping step (as in the two-phase Rankine cycle). Moreover, there must be provisions for the addition and rejection of heat.

The cycle may introduce the heat internally by combustion, as in the case for the well-known Otto and Diesel cycles and for the gas-fired turbine (the Brayton cycle). The combustants and combustion products constitute the working fluid, and the system is open.

Two other cycles that have been proposed and that remain of at least theoretical interest are the Stirling and Ericsson cycles. They are examples of closed cycles which would operate under limiting or hypothetical conditions, as is the Carnot cycle. The problems, of course, are in the implementation.

There is, in addition, what may be referred to as the nonadiabatic Joule cycle, which was introduced previously in Section 2.2 under the aegis of nonadiabatic differential compression and expansion. A more practical embodiment can be referred to as a Joule cycle using multistage adiabatic compression and expansion with interstage heat transfer—heat rejection in the compression section and heat addition in the expansion section. It is further detailed in Chapter 4.

A variation of the Rankine cycle is the adjustable proportion fluid mixture (APFM) cycle, which takes advantage of the change in temperature that occurs during the vaporization and condensation of a mixture, e.g., a binary mixture.

The preceding subjects are further discussed in the subsequent text for this section.

Stirling Cycle

The (closed) Stirling cycle uses a gaseous working fluid such as air. The sequential steps are as follows:

- Isothermal compression at a lower temperature, with heat removed
- Heat added at constant volume to reach a higher temperature
- Isothermal expansion at the higher temperature, with heat added
- Heat rejection at constant volume to reach the lower temperature

The Stirling cycle differs from the conventional Joule cycle in that compression and expansion are isothermal rather than adiabatic, and heat addition and rejection are at constant volume rather than constant pressure. The experimental difficulties have proven formidable.

Since the heat addition and rejection are at constant volume, they cancel out between the two common isothermal temperature levels. The efficiency is determined only from the isothermal expansion and compression at the two temperature levels. Thus

$$\text{Eff}_{\text{Stirling}} = \frac{RT_h \ln(V_2/V_1) - RT_c \ln(V_2/V_1)}{RT_h \ln(V_2/V_1)} = \frac{T_h - T_c}{T_h},$$

which is identical to the Carnot efficiency. The ratio V_2/V_1 denotes the compression—expansion ratio for a closed system.

Ericsson Cycle

The (closed) Ericsson cycle operates isothermally between pressure levels and consists of the following features. Heat is rejected only dduring compression and added only during expansion, which is somewhat analogous to the Stirling cycle. No additional heat is addedd or rejected, however. Because the expander exhaust gases are at a higher temperature than the compressed gases, a waste-heat exchanger is used between the two streams.

For isothermal compression and expansion, the ideal efficiency is

$$\text{Eff}_{\text{Ericsson}} = \frac{RT_h \ln(P_2/P_1) - RT_c \ln(P_2/P_1)}{RT_h \ln(P_2/P_1)} = \frac{T_h - T_c}{T_h},$$

where here again T_h and T_c represent the higher and lower temperatures maintained during expansion and compression and where P_2/P_1 is the compression–expansion ratio. Again, a theoretical efficiency is obtained which is equal to the Carnot efficiency.

Isothermal Compression and Isentropic Expansion

The working gas would be compressed isothermally, heated, and then expanded isentropically and adiabatically back to the original condition. (Alternately, or additionally, a waste-heat exchanger could be used if the expanded gases were at a temperature above the isothermal condition.) A *P-T* diagram is shown in Fig. 2.15.

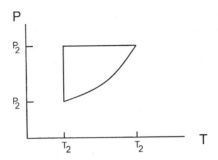

Fig. 2.15. Cycle with isothermal compression.

Operating between temperature levels T_h and T_c and at a compression–expansion ratio of P_2/P_1 as diagrammed, the efficiency would be

$$\text{Eff} = \frac{C_p(T_h - T_c) - RT_c \ln P_2/P_1}{C_p(T_h - T_c)}$$

$$= 1 - \frac{(R/C_p)\ln P_2/P_1}{T_h - T_c}$$

Since it is required that during an isentropic expansion $C_p\,dT = V\,dP$, then

$$\frac{C_p}{R}\ln\frac{T_h}{T_c} = \ln\frac{P_2}{P_1}$$

Therefore,

$$\text{Eff} = 1 - \frac{T_c \ln T_h/T_c}{T_h - T_c}.$$

An advantage is that circa ambient temperatures can be utilized for the isothermal compression or for a near-isothermal compression. Thus the fuel efficiency will be high, of the same order as for the Rankine cycle.

By comparisons, the approximate theoretical efficiency for the Rankine cycle is

$$\text{Eff}_{\text{Rankine}} = \frac{C_p(T_h - T_c)}{C_p(T_h - T_c) + \lambda} = 1 \Big/ \left(1 + \frac{\lambda}{C_p(T_h - T_c)}\right).$$

For the most part, the Rankine efficiency will be less than the efficiency for isothermal compression with isentropic expansion.

A representative comparison can be made as follows, say, based on steam, where $\lambda \sim 1000$ Btu/lb and $C_p \sim 0.5$ Btu/lb-°F. In turn, let $T_c = 100 + 460 = 560$°F and $T_h = 1120$°R. By substituting, the theoretical efficiency for an isothermal–isentropic cycle would be

$$\text{Eff} = 1 - \frac{560 \ln 2}{1120 - 560} = 30.7\%.$$

By comparison,

$$\text{Eff}_{\text{Rankine}} = 1 / \left(1 + \frac{1000}{0.5(560)} \right) = 21.9\%,$$

$$\text{Eff}_{\text{Carnot}} \sim \frac{1120 - 560}{1120} = 50.0\%.$$

(The Carnot efficiency is necessarily an approximation, since the conditions do not apply precisely.) There is a noticeable improvement for the isothermal–isentropic cycle efficiency over the theoretical Rankine efficiency, but it would be offset by compressor losses.

Observe that if the approximation for the logarithm is used,

$$\ln \frac{T_h}{T_c} \sim \frac{T_h/T_c - 1}{T_h/T_c} = \frac{T_h - T_c}{T_h}.$$

Then

$$\text{Eff} \sim 1 - \frac{T_c((T_h - T_c)/T_h)}{T_h - T_c} = 1 - \frac{T_c}{T_h} = \frac{T_h - T_c}{T_h},$$

which will be the same as the Carnot efficiency.

Note that the efficiency could be enhanced by transferring the heat of compression to, say, combustive air, where the combustants here are in turn fired countercurrently to the flow of the compressed working fluid. In the limit, if the exit temperature of the combustion products approaches the temperature of the working fluid after isothermal compression, then the theoretical efficiency could approach 100%. This is an idealization, but it indicates that there are ways to circumvent the limitations supposedly imposed by the Carnot cycle efficiency.

Joule Cycle with Continuous (or Differential) Nonadiabatic Compression and Expansion

The relationships for nonadiabatic behavior at varying temperature have been previously set forth in Section 2.2 for differential compression and

expansion. We are interested here in how these concepts can be combined
into a cycle, namely, a form or variant of the Joule cycle, which may be
referred to as a nonadiabatic Joule cycle.

The features of the nonadiabatic Joule cycle are detailed in Chapter 4
for incremental changes, that is, where heat transfer occurs between
stages. For the particular theoretical embodiment described here, heat
transfer occurs continuously during both compression and expansion. This
case may also be called nonadiabatic differential compression and expansion at varying temperature. (As a limiting condition, constant temperature
may be assumed.)

Moreover, as will be presented in Chapter 4 and Fig. 4.1, the heat
transfer from the working fluid can be made to combustive air during
compression, to which fuel is then added and combustion takes place. The
heat from the hot combustion product gases is then transferred to the
working fluid during expansion. The expanded working fluid can be further
cooled prior to recompression.

Alternately, as also will be shown in Chapter 4, a thermal or geothermal
source can serve in lieu of the combustive system and can in part be
recycled to the compression side of the cycle. In fact, a fluid (liquid) can be
heated externally by a combustive source (in countercurrent flow) and the
fluid in turn can serve as the heat transfer medium during compression
and expansion, with the fluid heat transfer medium moving concurrently
with the working fluid.

Some basic flow diagram options for differential nonadiabatic compression and expansion are indicated in Fig. 2.16. The corresponding cycles are
illustrated in Fig. 2.17 in P-T space. The legs T_a-T_b and T_c-T_d denote the
changes in temperature for the accompanying combustive or thermal
system.

As noted in Section 2.2, the cycle could operate if $T_4 = T_1$ and $T_3 > T_2$.
In P-T space the cycle would lie in the region to the left of the adiabatic
isentropic curve drawn through point $T_1 = T_4$, as denoted in Fig. 2.17(b);
that is, since

$$(C_p - f)\, dT = V\, dP = \frac{RT}{P}\, dP,$$

integrating at constant f yields

$$\ln\frac{P}{P_1} = \frac{C_p - f}{R}\ln\frac{T}{T_1}$$

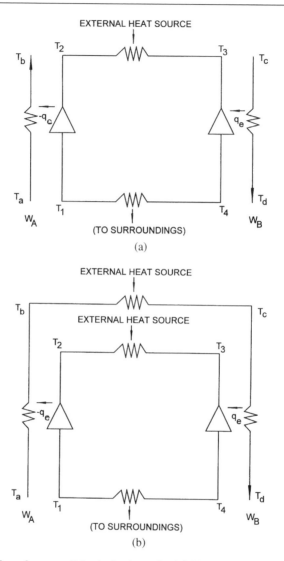

Fig. 2.16. Options for nonadiabatic Joule cycle. (a) Heat transfer during compression and expansion; (b) circulating heat recovery.

where $(-f)$ is positive for both compression and expansion, such that the corresponding slopes in log P-log T space will be greater than for the adiabatic isentropic curve drawn through point $T_1 = T_4$. Moreover, if $-f_{\text{comp}} > -f_{\text{exp}}$, the slope of the nonadiabatic compression locus will be greater than that for the nonadiabatic expansion locus. The temperature

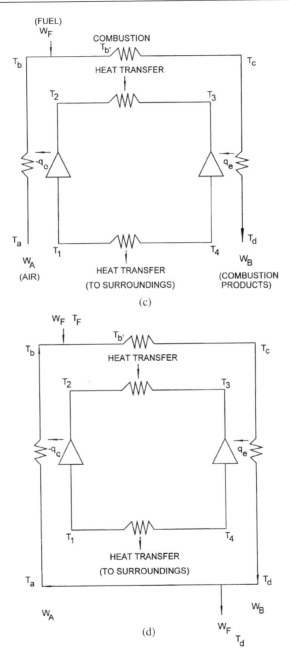

Fig. 2.16. (*Continued*) (c) fuel energy source; (d) thermal or geothermal energy source with recirculation.

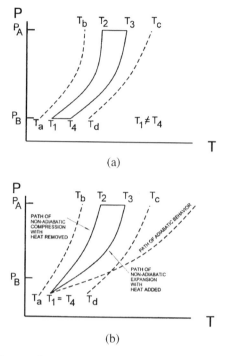

Fig. 2.17. *P-T* diagram for nonadiabatic Joule cycle. (a) $T_1 \neq T_4$; (b) $T_1 = T_4$.

difference, however, will be greater for the latter than for the former. It turns out that the work done by expansion,

$$w = (C_p - f_{\text{exp}})(T_3 - T_4), \qquad T_1 = T_4,$$

will be greater than the work required for compression,

$$(-w) = (C_p - f_{\text{comp}})(T_2 - T_1);$$

that is, the increased temperature difference for expansion more than offsets the decrease in $(C_p - f)$, as derived in Section 2.2 under the heading "Combined Compression and Expansion."

As a further limit, $T_4 = T_1$ *and* $T_3 = T_2$. In the case of the adiabatic Joule cycle, or Joule cycle proper, the temperature difference merely would cycle between the same upper and lower limits; there would be no heat transfer to or from the surroundings and there would be no net work done. In the nonadiabatic Joule cycle, for the same limiting conditions, but where the legs T_1-T_2 and T_3-T_4 are nonadiabatic, there still would be no net work done if the heat rejected equalled the heat added. The loci of compression and expansion can be viewed as identical.

As previously developed, however, in Section 2.2 under the heading "Special Case: Variable f_{comp} and f_{exp}," it is shown that net work can be done even if $T_4 = T_1$ and $T_2 = T_3$; that is, the behavior of the f-functions can be such that the locus of expansion will lie to the right of the locus of compression in P-T space. Another way of saying this is that the heat added during expansion will be greater than the heat rejected during compression.

Otherwise, $T_4 > T_1$ and/or $T_3 > T_2$, where nonadiabatic compression and expansion also are considered to be nonisothermal except as a limiting condition.

As an additional limit, for the circulating medium, if $T_d \rightarrow T_a$, then the overall theoretical cycle efficiency would approach 100%. This aspect also is examined in Chapter 4.

Closing the Heat Balance. For the several options shown in Fig. 2.16, the heat balances involving the heat transfer medium may be written as follows and may in turn be used to establish an overall efficiency rating.

The efficiency rating may be based entirely on the heat balances:

$$\text{Eff} = \frac{\text{heat added} - \text{heat lost}}{\text{heat added}} = 1 - \frac{\text{heat lost}}{\text{heat added}}.$$

For simplicity, the stream enthalpies may be assumed to have a constant heat capacity and a common reference enthalpy. The heat capacities for the medium are subscripted. The heat capacity for the working fluid is not subscripted.

The stream rates for the medium are based on a unit mass or mole of working fluid circulating.

OPTION (A)

The heat balances for the compression leg and expansion leg are

$$W_A(C_p)_A(T_b - T_a) = -q_c = -f_{comp}(T_2 - T_1),$$

$$W_B(C_p)_B(T_d - T_c) = -q_e = -f_{exp}(T_4 - T_3).$$

Therefore, for the overall efficiency,

$$\text{Eff} = 1 - \frac{-q_c + C_p(T_4 - T_1)}{q_e + C_p(T_3 - T_2)} = 1 - \frac{-f_{comp}(T_2 - T_1) + C_p(T_4 - T_1)}{f_{exp}(T_4 - T_3) + C_p(T_3 - T_2)},$$

where f_{comp} is negative and f_{exp} is negative. If $T_4 = T_1$, then the efficiency will be raised.

OPTION (B)

The heat balances again may be written as

$$W_A(C_p)_A(T_b - T_a) = -q_c = -f_{\text{comp}}(T_2 - T_1),$$

$$W_B(C_p)_B(T_d - T_c) = -q_e = -f_{\text{exp}}(T_4 - T_3),$$

where here $W_A = W_B$ and $(C_p)_A = (C_p)_B$. Furthermore, for the overall balance, in terms of the stream enthalpies, H,

$$W_B H_B - W_A H_A = W_A(C_p)_A(T_b - T_a) + W_A(C_p)_A(T_c - T_b)$$
$$+ W_A(C_p)_A(T_d - T_c)$$
$$= W_A(C_p)_A(T_d - T_a).$$

Therefore, for the efficiency,

$$\text{Eff} = 1 - \frac{W_A(C_p)_A(T_d - T_a) + C_p(T_4 - T_1)}{W_A(C_p)_A(T_c - T_b) + C_p(T_3 - T_2)}.$$

If $T_4 = T_1$, the efficiency increases. Then note, particularly, that if also $T_d = T_a$, the efficiency becomes 100%.

OPTION (C)

For combustion,

$$m(W_A + W_F) = W_B,$$

where the multiplier m is dictated by the combustion stoichiometry. For simplicity, however, let

$$W_B \sim W_A \quad \text{and also let} \quad (C_p)_B \sim (C_p)_A.$$

By the heat balance for combustion,

$$W_F \, \Delta H = W_B(C_p)_B(T_c - T_b),$$

where ΔH is the heat of combustion. Therefore, for the efficiency,

$$\text{Eff} = 1 - \frac{W_A(C_p)_A(T_d - T_a) + C_p(T_4 - T_1)}{W_F \, \Delta H}.$$

If $T_4 = T_1$ the efficiency is enhanced. If also $T_d = T_a$, then the efficiency becomes 100%.

OPTION (D)

Here, for the mixing of streams A and F,

$$W_A + W_F = W_B$$

and

$$W_A H_A + W_F H_F = W_B H_B$$

or

$$W_A(C_p)_A(T_b - T_0) + W_F(C_p)_F(T_F - T_0) = W_B(C_p)_B(T_b' - T_0),$$

where T_0 is the common reference temperature for the enthalpies. Assuming the stream heat capacities are equal, the foregoing expression reduces to

$$W_A(T_b) + W_F(T_F) = W_B(T_b').$$

The preceding relationship permits the determination of T_b' from the other stream temperatures. Therefore, for the efficiency,

$$\text{Eff} = 1 - \frac{W_B(C_p)_B(T_d - T_0) - W_A(C_p)_A(T_a - T_0) + C_p(T_4 - T_1)}{W_F(C_p)_F(T_F - T_d)}.$$

Since $W_B = W_A + W_F$, and assuming the heat capacities to be equal, the preceding equation reduces to

$$\text{Eff} = 1 - \frac{W_F(C_p)_F(T_d - T_0) + W_A(C_p)_A(T_d - T_a) + C_p(T_4 - T_1)}{W_F(C_p)_F(T_F - T_d)}.$$

If $T_4 = T_1$, then the efficiency is enhanced. If $T_d = T_a$ also, the efficiency is further increased to

$$\text{Eff} = 1 - \frac{T_d - T_0}{T_F - T_d}.$$

If the convention is adapted that $T_0 = T_d$, then the efficiency becomes 100%.

Multistage Adiabatic Compression and Expansion with Interstage Heat Transfer

The mathematics of the subject will be dealt with in Chapter 4. Here we will be concerned with the diagrammatic representation of a cycle as shown in Fig. 2.18. Heat is removed before and between the adiabatic compression stages and added before and between the adiabatic expansion stages, as indicated. The number of stages in the compression section is arbitrary and the number of stages in the expansion is also arbitrary. The number of stages in each section do not have to be the same either. For the purposes of illustration, however, three stages are shown in each section. In general, the last compressor stage is designated n and the last expansion stage is designated as m in the direction of flow.

There is a common lower-pressure level at the end of the expansion section and the beginning of the compression section, and a common higher-pressure level at the end of compression and the beginning of the

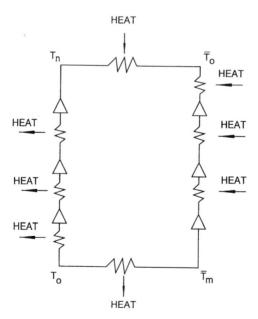

Fig. 2.18. Flow diagram for multistage adiabatic isentropic compression and expansion with interstage heat transfer and with heat rejection between $\bar{T}_m$ and T_0 ($\bar{T}_m > T_0$.

expansion section. Moreover, the pressure will be the same between successive stages. This said, there will, however, be small pressure-drop losses due to line and heat-exchanger frictional effects or other processing features. Note furthermore that the compression and expansion stages do not necessarily need to utilize the same corresponding values for the interstage pressures, or stagewise pressure changes, even if the same number of stages were used in each section. Moreover, the heat transferred in the successive interstage heat exchangers does not have to be uniform, but may vary from stage to stage.

Temperatures in the expansion section ($\overline{T}$) are distinguished from temperatures in the compressor section (T) by the addition of a bar above the symbol.

We will first be concerned with the more general case where $\overline{T}_m > T_0$ and $\overline{T}_0 > T_n$. This case is represented graphically in Fig. 2.19 for linear and logarithmic coordinates. As previously noted, the same number of stages do not have to be used in the compression section and the expansion section, nor does the degree of heat removal or addition have to be uniform. Only the overall pressure levels between compression and expansion must be the same. They are denoted as P_B for the upper pressure level and as P_A for the lower pressure level.

Alternatively, the cycle can be diagrammed as in Fig. 2.20, where here we are interested in the special case where $T_0 = \overline{T}_m$; the stagewise loci of compression with heat removal and of expansion with heat addition are indicated schematically in Fig. 2.21. As a further variation, the final expansion stage and the initial compression stage can theoretically both follow the same adiabatic isentropic path; that is, there would be no heat removal ahead of the first compressor stage. As expected, there will have

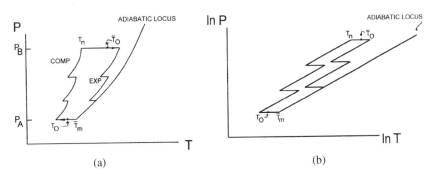

Fig. 2.19. Pressure versus temperature diagram for adiabatic isentropic compression and expansion with interstage heat transfer and with heat rejection between $\overline{T}_m$ and T_0 ($\overline{T}_m > T_0$). (a) P-T coordinates; (b) logarithmic coordinates.

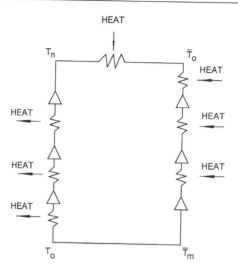

Fig. 2.20. Flow diagram for multistage adiabatic isentropic compression and expansion with interstage heat transfer and with no heat rejection between $\overline{T}_m$ and T_0 ($\overline{T}_m = T_0$).

to be an extra degree of heat load removal after, say, the first compression stage and/or prior to any or all of the other successive compression in order that the successive loci of compression remain to the left of the loci of expansion.

   Similarly, it can be stipulated that $T_n = \overline{T}_0$. As a further variation, both the final compression stage and the initial expansion stage theoretically can follow the same adiabatic isentropic path; that is, there would be no

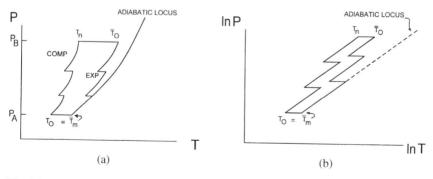

Fig. 2.21. Pressure versus temperature for multistage adiabatic isentropic compression and expansion with interstage heat transfer and with no heat rejection between $\overline{T}_m$ and T_0 ($\overline{T}_m = T_0$). (a) P-T coordinates; (b) logarithmic coordinates.

heat addition ahead of the first expansion stage. As expected, there will have to be an extra degree of heat addition after the first expansion stage and/or prior to any of the other successive expansion stages in order that the successive loci of expansion remain to the right of the loci of compression.

In log P-log T space, the paths or loci of the stagewise adiabatic isentropic compression and expansion would appear as straight lines as shown in Fig. 2.21(b). Furthermore, these paths or loci will tend to be parallel in log P-log T space. There is the consideration, however, that the same difference in log T at a higher temperature will produce a higher difference in T than at a lower temperature; that is, in the symbolism used, let

$$\log \overline{T}_0 - \log T_n = \log \overline{T}_m - \log T_0,$$

whereby it will follow that

$$\frac{\overline{T}_0}{T_n} = \frac{\overline{T}_m}{T_0} \quad \text{or} \quad \frac{\overline{T}_0}{T_n} - 1 = \frac{\overline{T}_m}{T_0} - 1.$$

Therefore,

$$\frac{\overline{T}_0 - T_n}{T_n} = \frac{\overline{T}_m - T_0}{T_0}$$

and

$$\frac{\overline{T}_0 - T_n}{\overline{T}_m - T_0} = \frac{T_n}{T_0} > 1;$$

that is, the heat added will be greater than the heat rejected and there must be net work produced.

It should be emphasized that greater horsepower will be delivered on the expansion side than will be required for the compression side. This imbalance will be reflected in the equipment sizes and ratings and in the volumetric capacities.

In the foregoing instances the heat from multistage compression can be recycled to the expansion side of the cycle, either via combustive air or by the recycling of part of a thermal or geothermal fluid medium, as has been shown previously for combined differential compression and expansion. The particulars are further detailed in Chapter 4 and are indicated here in Fig. 2.22.

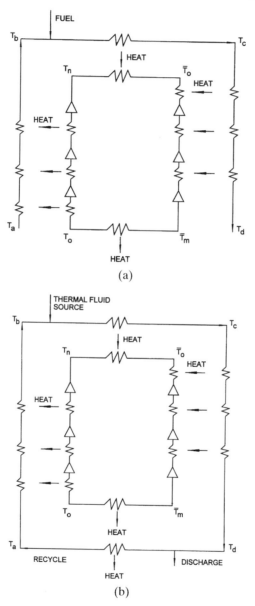

Fig. 2.22. Flow diagram for multistage compression and expansion cycle. (a) Air–fuel combustive energy source; (b) thermal or geothermal energy source.

Adjustable Proportion Fluid Mixture Cycle

The adjustable proportion fluid mixture (APFM) cycle takes advantage of the change in temperature which occurs in a mixture during vaporization and condensation in a Rankine cycle. A notable example is the Kalina cycle using a binary mixture of ammonia and water (3). The cycle may be analyzed as follows.

Consider first the temperature–composition diagram for a binary mixture A-B at constant pressure, where here A is the more volatile component. This scenario is illustrated schematically in Fig. 2.23. This case is sometimes called a T-x phase diagram in two-space, and it represents a constant-pressure section cut from a three-dimensional representation in P-T-x space (4).

Within the confines of the two-phase region or envelope, a mixture of total composition F will separate into a saturated vapor phase V, whose composition is given by the dew-point curve, and a saturated liquid L, whose composition is given by the bubble-point curve. The representation is called a "tie-line."

Furthermore, the ratio L/V is given by the ratio of the distance F-V to the distance L-F, which may be determined by the composition differences. Also the ratio L/F is given by the ratio of the distance F-V to the distance L-V, and the ratio V/F is given by the ratio of the distance L-F to the distance L-V; that is, by the material balances, for any component,

$$F = V + L,$$
$$Fx_F = Vy + Lx,$$

where F, V, and L are in mass or moles or in mass or moles per unit time as a flow rate, and the respective compositions x_F, y, and x are corre-

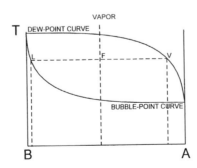

Fig. 2.23. Schematic temperature–composition diagram for a binary mixture at constant pressure.

spondingly in mass fraction or mole fraction. It follows then that

$$(V + L)x_F = Vy + Lx$$

and

$$\frac{L}{V} = \frac{y - x_F}{x_F - x}.$$

The other ratios can be determined similarly:

$$\frac{L}{F} = \frac{y - x_F}{y - x}, \qquad \frac{V}{F} = \frac{x_F - x}{y - x},$$

whereby

$$\frac{L}{F} + \frac{V}{F} = 1.$$

If so desired, loci of constant L/F or V/F ratios may be introduced, as indicated in Fig. 2.24.

The relationship between the composition of the vapor phase and the liquid phase, for a binary, is commonly known as the y-x curve, as pictured in Fig. 2.25. Its behavior is one of the concerns of the study of phase equilibria (4–6) and of its application to the vaporization and condensation of mixtures (7). It may be further observed that in the study of phase equilibria, compositions are usually expressed in mole fractions. The pattern of behavior is more uniform, or more regular, if the phase compositions are expressed in mole fractions.

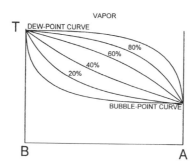

Fig. 2.24. Binary temperature–composition diagram showing constant percentages of vapor.

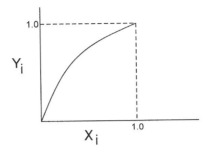

Fig. 2.25. Schematic y-x curve for a component i.

As a simplification, the y-x behavior may be represented as a straight line over limited domains of composition. Thus, for a component i and adding subscripts,

$$y_i = K_i x_i,$$

where K_i may behave as a constant, called the K-value or vapor–liquid equilibrium ratio for component i, and where y_i and x_i are the mole fractions of i in the vapor and liquid phases, respectively. Graphical representations for the K-values as functions of temperature and pressure have been made for the light hydrocarbons and a few nonhydrocarbons which occur in multicomponent petroleum and natural gas–gasoline mixtures (6), where the K-value behavior tends to be independent of composition. For binaries, these values are of limited use.

In other instances, the constant relative volatility relationship may be instituted such that, for the two components i and j,

$$\frac{y_i}{y_j} = \frac{K_i x_i}{K_j x_j} = \alpha_{i-j} \frac{x_i}{x_j},$$

where the relative volatility $\alpha = \alpha_{i-j}$ is a constant. Since the mole fractions in each phase sum to unity, it will follow that

$$\frac{y_i}{1 - y_i} = \alpha \frac{x_i}{1 - x_i},$$

whereby

$$y_i = \frac{\alpha x_i}{1 - x_i + \alpha x_i}.$$

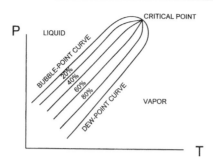

Fig. 2.26. *P-T* envelope for a mixture of constant total composition.

In the case of a gas of limited solubility, Henry's law may be invoked. This law relates the partial pressure of the gaseous component in the vapor phase to its concentration in the liquid phase such that

$$pp_i = Py_i = H_i x_i \quad \text{or} \quad y_i = \left(\frac{H_i}{P} \right) x_i,$$

where H_i is called Henry's constant and where it follows that $K_i = H_i/P$.

In *P-T* space, at constant composition, the two-phase envelope for a mixture of a given total composition may be represented schematically as in Fig. 2.26. The loci of constant liquid percentages are indicated.

Observe that vaporization proceeds from the bubble-point curve to the dew-point curve and that condensation proceeds inversely, from the dew-point curve to the bubble-point curve. Of special interest here are differential and stepwise condensation.

Differential and Stepwise Condensation. In differential or continuous condensation, with point withdrawal of the condensate, the path of the vapor composition during condensation will follow the dew-point curve toward the more volatile component as indicated by the arrow in Fig. 2.27, in the indicated sequence 5-4-3-2-1-0. The corresponding liquid-phase composition will follow the bubble-point curve, at the opposite end of the horizontal tie-lines. Note that the composition of the vapor phase changes, as does the composition of the condensate withdrawn.

For the point withdrawal of condensate, represented by $\delta L = dL$ with composition x_i, the differential material balance is

$$x_i \, \delta L = -d(Vy_i) = -V \, dy_i - y_i \, dV,$$

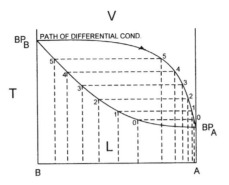

Fig. 2.27. Path of differential condensation on a temperature–composition diagram.

where

$$F = V + L \quad \text{and} \quad dV = -dL.$$

Therefore,

$$\frac{dL}{F - L} = -\frac{dy_i}{y_i - x_i},$$

where the size of F is arbitrary and may be assumed to be unity. Knowing the y_i-x_i behavior, the foregoing equality can be integrated to determine y_i versus L. At the same time, the behavior of x_i versus L will be established from the y_i-x_i curve. The temperature behavior will follow from the T-x_i diagram.

Most often, the lower limits will be $L = 0$ and $y_i = [y_0]_i = (x_F)_i$; that is, the initial value of the vapor-phase composition will be the designated composition of the mixture to be condensed.

The mean condensate composition will be

$$[x_i]_{\text{mean}} = \frac{\int_0^L x_i \, dL}{L}.$$

The mean composition of the vapor phase will, in turn, be

$$[y_i]_{\text{mean}} = \frac{F(x_F)_i - L[x_i]_{\text{mean}}}{F - L}.$$

Assuming constant heat capacity, the mean temperature of the condensate will be

$$[T]_{mean} = \frac{\int_0^L T\,dL}{L},$$

where T is represented as a function of L and may be established from the $T\text{-}x_i$ curve and the relationship between x_i and L as determined.

The condensate may be removed in discrete steps as also indicated in Fig. 2.27. The analysis would be the same provided that within each step there was pointwise withdrawal of the condensate.

If condensate flow is involved, the analysis becomes more complex. For the present situation, say, in concurrent flow,

$$F = V + L,$$
$$F(x_F)_i = Vy_i + Lx_i,$$

so that

$$dV = -dL,$$
$$0 = V\,dy_i + y_i\,dV + L\,dx_i + x_i\,dL.$$

Simultaneous numerical methods are necessary for the solution, with no separation of variables for the integration, starting at $L = 0$. It is necessary to know the $y_i\text{-}x_i$ behavior. The temperature behavior will follow dependently from the $T\text{-}x$ diagram.

If the condensate flow is countercurrent to the vapor flow, then difference-type material balances must be used:

$$V - L = V_0 - L_0 = F - L_0,$$
$$Vy_i - Lx_i = F(x_F)_i - L_0(x_0)_i,$$

where the subscript 0 denotes the vapor-phase inlet. By differentiating,

$$dV = dL,$$
$$V\,dy_i + y_i\,dV - L\,dx_i - x_i\,dL = 0,$$

Again, simultaneous numerical methods are required with the additional complication that it becomes trial-and-error for the condensate L_0 and its composition $(x_0)_i$ unless complete condensation takes place, whereby $L_0 = F$ and $(x_0)_i = (x_F)_i$.

(The foregoing expressions also will produce a contradiction for complete condensation in countercurrent flow, since it becomes necessary that $V = L$ and it will follow that $y_i = x_i$, which is contrary to the phase equilibrium behavior for a binary system.)

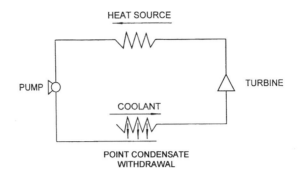

Fig. 2.28. APFM cycle with point condensate withdrawal.

For the foregoing reasons, it is preferable to consider only the pointwise withdrawal of condensate.

Adaptation to the Rankine Cycle. Differential condensation with point condensate withdrawal may be adapted to the Rankine cycle as shown in Fig. 2.28. Since the temperature will rise during vaporization, the heat source will be countercurrent to vaporization for more efficient heat transfer. Moreover, due to differential condensation, the mean temperature of the condensate will be intermediate between the boiling-point temperatures at the condensation pressure for the two components of the binary. This situation can serve to reduce the heat load. Since the temperature will be lowered during condensation, the coolant flow should be countercurrent to the condensate withdrawal sequence. Finally, condensation may proceed in steps as diagrammed in Fig. 2.29.

A schematic representation of the APFM cycle is shown in *P-T* coordinates in Fig. 2.30 for stepwise condensation. The two-phase envelope will change during condensation since the vapor-phase composition will be changing during condensation.

Factors that affect the efficiency, and which may be partially offsetting, are the latent heat changes during vaporization and condensation versus the work delivered from the expansion of the mixture.

Countercurrent Heat Transfer. An advantage of the APFM or Kalina cycle is that for a binary (or multicomponent) working fluid, the temperature during vaporization varies (increases) and the temperature during condensation also varies (decreases). Using countercurrent flow between the hot side and cold side of the heat-exchanger surfaces, this temperature variation for the working fluid will produce a larger ΔT or thermal driving force across the tubewalls during both vaporization and condensation—

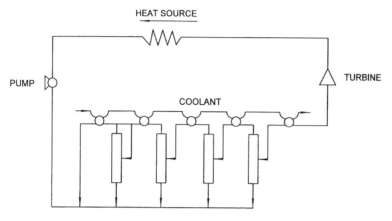

Fig. 2.29. APFM cycle with stepwise condensation.

other things being equal. The net effect can be construed as ensuring a greater heat transfer flux during both vaporization and condensation.

2.8. HEAT PUMP AND REFRIGERATION CYCLES

Refrigeration or heat pump cycles involve the single-phase expansion or two-phase expansion of a working fluid, whereby a cooling of the fluid medium takes place.

Single-phase expansion usually involves only the Joule–Thomson effect, or throttling effect, although expansion may also take place through a turbine. The system then can also be considered a heat engine.

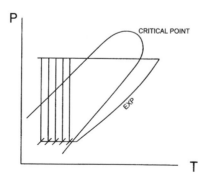

Fig. 2.30. Schematic APFM cycle with differential condensation. Note that the *P-T* diagram will change during compression since the vapor-phase composition changes.

Two-phase expansion—say, starting with a liquid phase—involves latent heat effects via vaporization. More pronounced cooling can take place and may be enhanced by Joule–Thomson effects occurring in the gaseous phase during the phase change.

The cooled working fluid is partially heated by exchange with the localized surroundings and then is further heated during recompression to the high-side pressure. This step is followed by cooling (and condensation) at pressure by heat exchange with other localized surroundings (at a different temperature level). The reexpansion completes the cycle.

Whether the system is called refrigeration or a heat pump depends upon the objective. Thus the former emphasizes cooling the (localized) surroundings, whereas the latter emphasizes the heating of the (localized) surroundings. The operation may, in fact, be reversed by rerouting the flow to and from the heat transfer sink (the surroundings), e.g., for an in-home cooling–heating system. An optimum cochoice of refrigerant versus temperature and pressure levels is involved. More will be said on the subject in Chapter 5 on latent heat recovery. Of prime consideration is the cooling or heating effect as compared to the pressure–volume work of compression in terms of a coefficient of performance.

A departure from mechanical refrigeration is the absorption cycle whereby the exhaust vapors from expansion are absorbed in a cooled liquid medium. The resulting mix is pumped to pressure and heated. The working fluid component(s) is vaporized or distilled from the mixture to be cooled and/or condensed prior to expansion. The unvaporized phase is cooled and recycled to the absorber to serve as the absorbing medium. Examples are the ammonia–water system and the sulfur dioxide–water system. The pure components, ammonia and sulfur dioxide, also serve as working fluids in mechanical refrigeration cycles. Another option is carbon dioxide, which may be used in absorption refrigeration with a suitable alkaline medium that will permit the regeneration of the carbon dioxide, e.g., potassium carbonate solutions or ethanolamine solutions. Carbon dioxide also can be used as the working fluid in mechanical refrigeration.

Another variant is the Platen–Munters continuous absorption system, such as is used in the Electrolux refrigerator. A third component, which is a noncondensable gas (hydrogen), is added to the ammonia–water system. Vaporization is thereby enhanced, with less pressure-drop. A notable feature is that the vaporized product is (partially) condensed at a higher elevation or position, thus providing a liquid leg to maintain circulation. Only the heat added at the vaporizer is needed, therefore, to maintain circulation. The required juxtaposition of phase compositions and pressure and temperature levels is remarkable.

Coefficient of Performance

The effectiveness of a mechanical heat pump or refrigeration cycle is customarily measured in terms of a coefficient of performance (COP), which is the ratio of the heat delivered from or to the system divided by the net energy for compression.

The simplest embodiment of a heat pump cycle is the compression of a single-phase gaseous working fluid, followed by the cooling of the working fluid and then a Joule–Thomson expansion to its original state. (For the Joule–Thomson expansion of a perfect gas, the temperature does not change, as is approximately the case for liquids, which may have a small increase.) The cooling of the working fluid denotes heat delivered in whole or in part. At best, the COP would be unity, but will depend upon efficiency losses. The COP for refrigeration can be determined similarly. (For latent heat pumps, the COP can be greater than unity by virtue of the way it is defined.)

The reverse or inverse Joule cycle can be employed also. (The Joule power cycle is developed at length in the next chapter.) Here, the maximum COP will be the inverse of the Joule efficiency; in this case, the COP is $\alpha/(\alpha - 1)$. The COP for the corresponding refrigeration cycle would be the heat pump COP divided by α, which gives a value of $1/(\alpha - 1)$. The quantity α is determined as the ratio of the absolute temperatures before and after expansion or after and before compression.

More often, a two-phase working fluid is employed; this is the reverse or inverse of the Rankine cycle. Rigorously, the heat pump COP would be the reciprocal of the Rankine efficiency, and the COP for refrigeration would be apportioned from this value (based on the latent heat change as compared to the total enthalpic change at the higher pressure). Instead, however, a simplification is made based only on the latent heat change. Thus

$$\text{COP} \sim \frac{\Delta H_{\text{WF}}}{C_p(T_S - T_R)},$$

where $\Delta H_{\text{WF}} = \lambda_{\text{WF}}$ is the latent heat of vaporization for the working fluid. The temperatures T_S and T_R represent the maximum and minimum temperatures attained by the working fluid at the saturation curve. The upper temperature T_S denotes condensation; the lower temperature T_R denotes vaporization of the working fluid—the opposite of the Rankine cycle, as diagrammed previously in Fig. 2.5.

Using ambient conditions as the source and sink, and assuming at 20°F approach at each level, the difference $T_S - T_R$ would be circa 40°F. This difference will determine the maximum COP that can be attained for both a heat pump and a refrigeration cycle.

The foregoing reverse Rankine cycle also can serve as a latent heat pump. The working fluid vaporizing at T_R will absorb the heat of condensation from a second system, and reject it at T_S and at T_S the latent heat of condensation of the working fluid will supply the latent heat of vaporization for another, or third, system.

In the case of drying, the temperature level of a second system will remain essentially constant throughout for both vaporization and condensation. In this instance, a temperature difference of circa 40°F for the working fluid between T_R and T_S will serve, which permits a correspondingly high value for the COP, as will be shown in Chapter 5.

REFERENCES

1. Keenan, J. H. and F. G. Keyes, *Thermodynamic Properties of Steam*, Wiley, New York, 1936.
2. Hougen, O. A., K. M. Watson, and R. A. Ragatz, *Chemical Process Principles. Part II. Thermodynamics*, Wiley, New York, 1947, 1959.
3. Marsten, C. H., "Development of the Adjustable Proportion Fluid Mixture Cycle," *Mech. Eng.*, September (1992).
4. Hoffman, E. J., *Azeotropic and Extractive Distillation*, Interscience, New York, 1964. (Reprinted by Krieger, Huntington, NY, 1977.)
5. Hoffman, E. J., *The Concept of Energy: An Inquiry into Origins and Applications*, Ann Arbor Science, Ann Arbor, MI, 1977.
6. Hoffman, E. J., *Phase and Flow Behavior in Petroleum Production*, Energon, Laramie, WY, 1981.
7. Hoffman, E. J., *Heat Transfer Rate Analysis*, PennWell, Tulsa, OK, 1980.

CHAPTER 3

Heat Engines
and the Joule Cycle

The joule cycle involves the successive compression, heating, expansion, and cooling of a gaseous working fluid. The gaseous system is generally considered closed, and may be emphasized as a closed Joule cycle. If the gaseous system is open, then the cycle is called a Brayton cycle. Alternately, the Joule cycle can be called a closed Brayton cycle and the Brayton cycle called an open Joule cycle. The terms are somewhat interchangeable. The generic term is "heat engine," although "air engine" also has been used when the working fluid is air. The performance of the heat engine is limited by the temperature at which heat can be rejected to the surroundings and by the exit temperature of the combustion products or stack gases that furnish heat.

In the Joule cycle proper, the working fluid is heated by indirect heat transfer with a heat source, which is usually a combustion system that effects higher temperatures for the working fluid, which increases the efficiency. Nevertheless, the efficiency of the Joule cycle is most often something to be desired, except under unusual circumstances, for example, rejecting the waste-heat to outer space. However, with the adaptations of nonadiabatic compression and expansion as described in Chapter 2 (which will be expanded upon in Chapter 4), the Joule cycle can be modified to achieve high overall efficiencies for the conversion of heat to work.

The Brayton cycle involves equivalent modes of compression, heating, expansion, and cooling, but employs an open system as indicated. It is the cycle used for gas-fired turbine/generators. More specifically, heating may be accomplished by direct combustion after the compression step, where the combustibles and air and then their combustion products form the working fluid. The cooling or heat rejection step as such is not required, but may be used instead to preheat the combustants, for example, the air or air–fuel mixture used. The latter step is referred to as regenerative heat transfer, and it changes the performance criteria.

Whereas the closed Joule cycle can utilize dirty or solid fuels as the heat source, the open Brayton cycle requires clean-burning fuels in order to minimize corrosion and erosion of the turbine-expander blades.

The Joule cycle is analyzed in the subsequent pages as to performance criteria in terms of temperature and pressure, and representative working fluids. The means to improve the efficiency by regenerative heat transfer are presented and the effects of additional stages of compression are discussed. Any attempt to use a heat pump or reverse Joule cycle to transmit the heat rejected to the heat added will be nonproductive. The combustion system that supports the cycle can itself be used to upgrade the overall efficiency, however, by using the waste-heat rejected from the cycle to preheat the combustants. A continuation of these developments leads to a reconsideration of nonadiabatic compression and expansion, and the Joule cycle is examined in these terms.

Although we may speak of the efficiency of the Joule cycle per se, the more inclusive measure is the overall efficiency, which takes into account the combustive or thermal support as well. Accordingly, heat balances and overall efficiencies are derived on this basis. The in-depth extension is to the applications in Chapter 4.

3.1. COMPRESSION AND EXPANSION

The enthalpic behavior of a gas during compression and expansion is based on the following identities, as previously developed:

$$dH = C_p \, dT - C_p \mu \, dP = V \, dP + d\omega,$$

where, in consistent units,

$$H = \text{enthalpy function},$$
$$C_p = \text{heat capacity (molar)},$$
$$\mu = \text{Joule-Thomson coefficient},$$
$$T = \text{temperature in degrees absolute},$$
$$P = \text{pressure},$$
$$V = \text{volume (molar)},$$
$$\omega = \text{intrinsic energy}.$$

Since

$$d\omega = d\text{lw} = T \, dS,$$

where

$$\text{lw} = \text{lost work},$$
$$S = \text{entropy},$$

then for an isentropic change, $dS = 0$ and $d\omega = d\text{lw} = 0$. Moreover, for a perfect gas, $\mu = 0$. It can be argued furthermore—as indicated in Section 1.8—that the descriptor "isentropic" says it all. In other words, the way in which it is employed also implies adiabaticity; that is, if $d\omega = 0$, then for a perfect gas,

$$dq - dw = C_p\,dT = V\,dP$$

or

$$-dw = C_p\,dT - dq = V\,dP - dq$$

and

$$C_p\,dT = V\,dP,$$

which will yield the P-T locus for an isentropic compression or expansion. Thus the behavior of dq or say $dq/dT = f$ will have no effect on the P-T locus. There is no built-in provision for the effect of dq or its derivatives on the P-T locus, and the result would be the same as if $dq = 0$. There will, however, be an effect on the behavior of the work term, that is, of

$$dH = C_p\,dT - \frac{dq}{dT}\,dT = C_p\,dT - f\,dT = V\,dP + d\omega.$$

For the foregoing reasons, the concept of a generalized irreversibility change $d\omega^*$ was introduced to describe nonadiabatic compression and expansion as presented in Section 2.2. Here, if it is required that the generalized irreversibility $d\omega^* = (d\omega - dq) = 0$, such that

$$-dw = (C_p\,dT - f)\,dT = V\,dP,$$

then the P-T locus can be adjusted for nonadiabatic behavior.

The aforestated simplifications lead to the customary integrated T-P relationship for the isentropic compression or expansion of an ideal gas:

$$\frac{T_2}{T_1} = \left(\frac{P_2}{P_1}\right)^{(k-1)/k} = \alpha,$$

where $k = C_p/C_v$ and P_2/P_1 is the compression or expansion ratio.

Furthermore, the overall heat balance requires that, in heat units,

$$dQ - dW = dH,$$

where

$$Q = \text{heat added to the system,}$$

$$W = \text{work done by the system (in heat units).}$$

If the compression or expansion is also adiabatic ($dQ = 0$), then for an ideal gas,

$$-dW = dH = C_p \, dT,$$

and at constant heat capacity, integrating between limits,

$$-W = -\Delta W = C_p(T_2 - T_1).$$

The convention used is that for work done on the system (compression), W will have a negative value; for work done by the system (expansion), W will have a positive value.

For a real gas, not only should the Joule–Thomson coefficient μ be recognized, as well as the intrinsic energy change or lost work, but the compressibility factor z in the equation of state $PV = zRT$ should be taken into account. It is, however, more convenient to incorporate these nonidealities into an overall efficiency rating.

Rearrangements

Utilizing the substitution

$$\frac{T_2}{T_1} = \alpha = \left(\frac{P_2}{P_1}\right)^{(k-1)/k},$$

it follows that, per unit mole or mass of working fluid,

$$-W = C_p T_1(\alpha - 1)$$

$$= C_p T_2\left(1 - \frac{1}{\alpha}\right) = C_p T_2 \frac{\alpha - 1}{\alpha}.$$

This permits a pronounced convenience in the derivations.

3.2. GENERALIZED CYCLE

The Joule cycle is represented schematically as in Fig. 3.1. The subscripted absolute temperature T is used to designate the successive stages. Further information is provided, for instance, in references 1, 2, and 3.

It is required that

$$\frac{T_2}{T_1} = \alpha, \qquad \frac{T_3}{T_4} = \alpha,$$

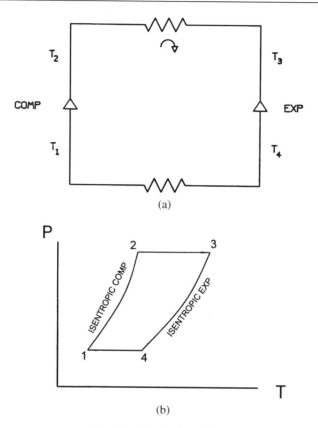

Fig. 3.1. The Joule cycle.

since $P_2 = P_3$ and $P_1 = P_4$. Flow of the working fluid is clockwise in the drawing.

The net work done or delivered in heat units, per unit of working fluid, is

$$W_{net} = C_p(T_3 - T_4) - C_p(T_2 - T_1)$$
$$= C_p(T_3 - T_2) - C_p(T_4 - T_1).$$

Alternately,

$$W_{net} = C_p T_3(\alpha - 1) - C_p T_2(\alpha - 1)$$
$$= C_p(\alpha - 1)(T_3 - T_2).$$

If $T_3 > T_2$, the cycle serves as a heat engine; if $T_3 < T_2$, the cycle serves as a heat pump, which requires a net heat input.

3.3. EFFICIENCY OF THE JOULE CYCLE

The efficiency of the Joule cycle may be written simply as

$$\text{Eff}_{\text{Joule}} = \frac{C_p(T_3 - T_4) - C_p(T_2 - T_1)}{C_p(T_3 - T_2)},$$

where the denominator represents the heat input delivered to the cycle (as distinguished from the heat of combustion of the fuel used). In terms of α,

$$\text{Eff}_{\text{Joule}} = \frac{T_3(1 - 1/\alpha) - T_2(1 - 1/\alpha)}{T_3 - T_2} = 1 - \frac{1}{\alpha}.$$

Since $\alpha = T_2/T_1 = T_3/T_4$, then

$$\text{Eff}_{\text{Joule}} = \frac{T_2 - T_1}{T_2} = \frac{T_3 - T_4}{T_3}.$$

These expressions are theoretical and are based on the heat delivered to the system rather than on the heat of combustion of the fuel used. The latter also will involve efficiency losses; thus, the overall efficiency based on the fuel will be less than the theoretical efficiency. The efficiency losses during compression and expansion will further reduce the efficiency.

The value of α will be limited by the temperature T_1 at which heat can be rejected to the surroundings and by the temperature T_2 that corresponds to the exit stack-gas temperature for the fuel combustion products (in countercurrent flow).

Using air as an example,

$$k = 1.4, \qquad \frac{k-1}{k} = 0.28,$$

whereby

P_2/P_1	α	Eff
1	1	0.0
2	1.214	17.6
5	1.569	36.3
10	1.905	47.5
100	3.630	72.5

The preceding comparison is an idealization, and due to temperature limitations as noted, it may not be possible to realize these compression ratios in practice, except, say, perhaps in outer space, where there are no restrictions on heat rejection.

Topping Cycle

Another possibility, of course, would be to utilize the Joule cycle as a "topping" cycle, whereby the (hot) exit gases from combustive support could be used for other purposes. These other purposes could include a secondary power generation cycle, as with the Rankine cycle (sometimes called the binary cycle concept), or else use as waste-heat for steam generation, space heating, and so forth. Enter the subject of cogeneration.

Optimum Joule Cycle

Given the two extremes in temperature, T_3 and T_1, the object is to determine the optimum intermediate values T_4 and T_2. The net energy output for the Joule cycle per unit of working fluid may be represented as

$$\text{net energy} = C_p(T_3 - T_4) - C_p(T_2 - T_1)$$

$$= C_p T_3\left(1 - \frac{1}{\alpha}\right) - C_p T_1(\alpha - 1).$$

The object then becomes to determine the optimum value for α, assuming the other terms are constants. Differentiating with respect to α and setting the derivative equal to zero,

$$0 = \frac{T_3}{\alpha^2} - T_1,$$

whereby

$$\alpha^2 = \frac{T_3}{T_1} \quad \text{or} \quad \alpha = \left(\frac{T_3}{T_1}\right)^{1/2}.$$

Since

$$\alpha = \frac{T_3}{T_4} = \frac{T_2}{T_1},$$

then

$$\alpha^2 = \frac{T_3}{T_4} \cdot \frac{T_2}{T_1}.$$

The optimum condition will be met if $T_2 = T_4$. That is, since

$$\alpha^2 = \frac{T_3}{T_1} \cdot \frac{T_2}{T_4} \quad \text{or} \quad \alpha = \left(\frac{T_3}{T_1}\right)^{1/2} \cdot \left(\frac{T_2}{T_4}\right)^{1/2}$$

substituting in the expression for net energy will yield

$$\text{net energy} = C_p T_3 \left[1 - \frac{1}{(T_3/T_1)^{1/2}(T_2/T_4)^{1/2}}\right]$$
$$- C_p T_1 \left[(T_3/T_1)^{1/2}(T_2/T_4)^{1/2} - 1\right]$$

Differentiating with respect to (T_2/T_4) as the variable while holding (T_3/T_1) constant and setting the derivative equal to zero, will, after clearing fractions, yield

$$\frac{T_2}{T_4} = 1$$

It may be noted that in the Rankine cycle the condition occurs whereby $T_4 \sim T_2 \sim (\sim T_1)$.

Efficiency. The efficiency can in turn be expressed as

$$\text{Eff}_{\text{Joule}} = 1 - \frac{1}{\alpha} = 1 - \left(\frac{T_1}{T_3}\right)^{1/2}.$$

As an illustrative example, if $T_1 = 100 + 460 = 560°R$ and $T_3 = 800 + 460 = 1260°R$, then

$$T_2 = T_4 = \sqrt{(560)(1260)} = 840°R \text{ or } 380°F$$

and

$$\text{Eff}_{\text{Joule}} = 1 - \sqrt{560/1260} = 27.6\%.$$

The high value for T_2 indicates that heat regeneration should be used on the combustion products or stack gases to counteract fuel efficiency losses.

As another example, if T_3 is raised to $T_3 = 1400 + 460 = 1860°R$, then

$$T_2 = T_4 = \sqrt{(560)(1260)} = 1021°R \text{ or } 561°F$$

and

$$\text{Eff}_{\text{Joule}} = 1 - \sqrt{560/1860} = 45.1\%$$

The use of heat regeneration between the combustion products and the fuel–air mixture or combustive air becomes even more important.

At the same time, of course, it would be most opportune if the Joule cycle efficiency itself could be increased, say, by returning waste-heat to the forefront of the cycle—if this were possible.

Rankine Cycle

In the Rankine cycle, by comparison, expansion is followed by cooling and condensation, and then the liquid condensate is pumped back to pressure to be vaporized and superheated. Operation is generally away from the vapor-pressure curve or two-phase envelope (the latter for a working fluid of two or more components). The pumping requirement is relatively negligible, so in effect the compression losses in the Joule cycle are traded for heat losses by condensation.

The Rankine efficiency is expressed more rigorously in terms of enthalpic changes (3). As an approximation, however, expression may be made in terms of the Joule cycle nomenclature as

$$\text{Eff}_{\text{Rankine}} = \frac{C_p(T_3 - T_4)}{C_p(T_3 - T_4) + \Delta H},$$

where $\Delta H = \Delta H_v = \lambda$ is the latent heat of vaporization and where $T_4 \sim T_1 \sim T_2$. Therefore,

$$\text{Eff}_{\text{Rankine}} = \frac{C_p T_3(1 - 1/\alpha)}{C_p T_3(1 - 1/\alpha) + \Delta H}$$

$$= \frac{\text{Eff}_{\text{Joule}}}{\text{Eff}_{\text{Joule}} + \Delta H/C_p T_3}.$$

In principle—other things being equal—the efficiency of the Joule cycle is greater than that for the Rankine cycle. Alternately,

$$\text{Eff}_{\text{Joule}} = \frac{\Delta H}{C_p T_3} \frac{\text{Eff}_{\text{Rankine}}}{1 - \text{Eff}_{\text{Rankine}}};$$

that is, if $\Delta H/C_p T_3 > 1$, then the Joule cycle would be more efficient. More precisely, if $\Delta H/C_p T_3 > (1 - \text{Eff}_{\text{Rankine}})$, then the Joule cycle would be more efficient.

In the Rankine cycle, sensible heat effects also are incurred in the liquid phase and in the vapor phase, for example, in supercooling the condensate and superheating the vapor. Moreover, if multicomponent working fluids are used, there will be a temperature change during condensation and vaporization. The exception to the latter is with azeotropes, which are constant boiling (or condensing) mixtures of a specific composition.

Carnot Cycle

Using the same nomenclature, the Carnot efficiency would be

$$\text{Eff}_{\text{Carnot}} = \frac{T_3 - T_4}{T_3}$$

if $T_1 \sim T_4$; that is, heat is rejected at some common lower temperature designated as T_4. Accordingly,

$$\text{Eff}_{\text{Carnot}} = \frac{T_3(1 - 1/\alpha)}{T_3} = 1 - \frac{1}{\alpha}.$$

Therefore, hypothetically at least, the Joule cycle and Carnot cycle efficiencies are comparable.

Note that in the Carnot cycle heat is added and rejected at constant temperature rather than at constant pressure as in the Joule cycle. There is not a direct analogy, therefore, between the two.

Brayton Cycle

The Brayton cycle delivers heat directly to the turbine in the form of combustion gases; the off-gases then are rejected to the atmosphere. Ideally, the corresponding efficiency would be

$$\text{Eff}_{\text{Brayton}} = \frac{C_p(T_3 - T_4) - C_p(T_2 - T_1)}{C_p(T_3 - T_2)}$$

$$= \frac{C_p T_3(1 - 1/\alpha) - C_p T_2(1 - 1/\alpha)}{C_p(T_3 - T_2)}$$

$$= 1 - \frac{1}{\alpha},$$

which is theoretically the same as for the Joule cycle.

The Brayton cycle has the advantage, however, that the heat rejection temperature level can be based on T_4 rather than T_1, and, in fact, T_4 conceivably can be at a level lower than the temperature of the surroundings. Moreover, there is no restriction on T_2, so that larger values of α can be used. Additionally, heat regeneration may be used between the combustion product off-gases and the combustive air.

Heat Regeneration. If heat regeneration is used between T_2 and T_4, then in the limit $T_2 \sim T_4$. The efficiency then becomes, ideally,

$$
\begin{aligned}
\mathrm{Eff}_{\mathrm{regen}} &= \frac{C_p(T_3 - T_4) - C_p(T_2 - T_1)}{C_p(T_3 - T_2)} \\
&= 1 - \frac{T_4 - T_1}{T_3 - T_2} \\
&= 1 - \frac{T_2 - T_1}{T_3 - T_2} = 1 - \frac{T_2(1 - 1/\alpha)}{T_3 - T_2} \\
&= 1 - \frac{T_1(\alpha - 1)}{T_3 - T_2}.
\end{aligned}
$$

Note that as α decreases, the efficiency increases. Thus the efficiency is favored by lower compression ratios and by higher values of T_3.

Hypothetically, as $\alpha \to 1$, the theoretical efficiency becomes unity, that is, 100%. This idealization is, however, subject to such limitations as regenerative heat-exchanger size and compression and expansion efficiency losses. In all other cases, the efficiency is limited by the temperature $T_2 \sim T_4$, which corresponds to the exhaust temperature of the combustion product gases.

Observe, furthermore, that for high efficiencies it will be required that

$$ T_3 - T_2 \sim T_4 - T_1 $$

or

$$ T_3 - T_2 \sim T_2 - T_1. $$

The implication is that, for low compression ratios, T_2 approaches T_1. Hence the difference $(T_3 - T_2)$ will become increasingly small. Thus the energy input and net work done also will fall off.

The difficulty, therefore, is that the net energy produced per unit of circulating working fluid will decrease, whereby mechanical efficiency losses, including those from compression and expansion, become controlling.

3.4. WORKING FLUIDS

A theoretical comparison for a number of representative working fluids at standard conditions is provided in Table 3.1, based on information in the literature, for example, references 4 and 5. For the most part, any gas could suffice, assuming the operation is in the gaseous region well away from the saturation curve and the critical region.

Presumably, the higher the k-value, the more desirable the substance is as a working fluid. Helium, therefore, is exemplary. Helium–xenon mixtures also have been used for studies on the Joule cycle, for example, for space applications. As to Freons (or freons), although the k-values may appear to be low, they have the desirable property of "paralleling" the saturation curve during compression and expansion, rather than tending to intersect. A disadvantage, much publicized, is that Freons are suspected of adversely affecting the ozone layer. The leakage factor may, therefore, become controlling.

In early investigations of the Joule cycle, air was used as the working fluid; hence the Joule cycle was called an air engine.

Table 3.1. PROPERTIES OF WORKING FLUIDS

	MW	C_p	k	$(k-1)/k$	T_c (°F)	P_c (psia)	P_{sat} (psia) 120°F	800°F
Air	29	6.96	1.40	0.28	−221.3	547	> Crit	> Crit
Nitrogen	28	6.95	1.40	0.28	−232.4	493.0	> Crit	> Crit
Hydrogen	2	6.816	1.41	0.29	−399.8	188.1	> Crit	> Crit
Helium	4	4.706	1.73	0.42	−456.6	33.2	> Crit	> Crit
Carbon dioxide	44	8.76	1.29	0.17	87.9	1071	> Crit	> Crit
Methane	16	8.426	1.309	0.24	−116.3	667.8	> Crit	> Crit
Ethane	30	12.291	1.193	0.16	90.09	707.8	> Crit	> Crit
Propane	44	17.076	1.132	0.12	206.01	616.3	240	> Crit
i-Butane	58	22.478	1.097	0.088	274.98	529.1	100	> Crit
n-Butane	58	22.429	1.097	0.088	305.65	550.7	70	> Crit
n-Pentane	72	27.842	1.077	0.071	385.7	488.6	22	> Crit
Freon-11	137.35	21.564	1.101	0.092	—	—	33.2	—
Freon-12	120.9	18.135	1.123	0.11	233.6	600	171.8	> Crit
Freon-21	102.9	14.612	1.157	0.14	—	—	55.75	—
Dowtherm-A	165	52.8	1.039	0.038	—	—	0.2	> 150

For the Rankine cycle and its inverse—refrigeration—it is necessary that the working fluid encounter the two-phase region. Here, working fluids such as Freons also are used because they possess the necessary two-phase characteristics at reasonable temperatures and pressures.

For the effect of temperature and pressure on the Joule cycle, generally speaking, the coefficient k will increase with decreasing temperature and increase with increasing pressure. This effect can be more complicated at lower temperatures, however, where for nitrogen the ratio or difference of the heat capacities will exhibit maxima (see, for instance, reference 4, pp. 232–233). The maxima appear in the isotherms.

3.5. BOUNDARY CONDITIONS

The performance of the Joule cycle proper is linked to temperature levels and pressure ratios and to the nature or properties of the working fluid, which remains in the single-phase gaseous region.

In general, gases with higher k-values are indicated, with the qualification that k tends to increase with increasing pressure but falls off with increasing temperature. The former property, however, usually offsets the latter. The lower hydrocarbons and the Freons, which have relatively low k-values at standard conditions, can nevertheless be used successfully as working fluids provided the thermal stability is not breached.

The extreme upper temperature levels for inert gases are dictated by the materials of construction, for both the heat transfer surfaces and for the turbine-expander. Although the turbine-expander blades may rise to temperatures approaching 2000°F (3) or even higher with more modern materials, the heat transfer surfaces are generally limited to circa 1400°F, which is the upper temperature level used in natural gas steam-reforming, for instance.

Although turbine-expanders can operate over wide ranges of pressure ratios, the compression step is more restrictive. Ordinarily, compression ratios are limited to circa 5/1 for each stage (3). Higher ratios will require multistage compression, usually with intercooling. Intercooling, however, will assault the power cycle efficiency. Without intercooling, the compressor outlet temperature should probably be kept at 800°F or lower, although higher temperatures are a matter of course for the compression stroke in a diesel engine (which uses compression ratios of around 17/1).

Another reason to keep temperature levels at circa 800°F or lower is to ensure thermal stability in the case of organically derived working fluids; that is, organic working fluids will start to decompose or "crack" at these temperature levels. This predisposition to break down increases with molecular weight; that is, methane is the most stable. (The presence of moisture, however, will accelerate decomposition via a steam-reforming action.) For such a situation, a separation step to remove decomposition products may be required if these products adversely affect the cycle.

Temperature Levels. Chief among the many other considerations are the temperature levels for heat rejection to the surroundings and for heat recovery from the combustion product gases. The former limits T_1 to about 100°F, the latter limits T_2 to about 200°F. Therefore,

$$\alpha \sim \frac{200 + 460}{100 + 460} = 1.18,$$

which severely limits the potential of the Joule cycle.

Although the use of a heat pump cycle would permit T_1 to be at a lower temperature and, in turn, reject the waste heat to the surroundings at a higher temperature, such a use is counterproductive. Even theoretically, there is no advantage, because the overall efficiency would remain the same. In reality, with the extra accoutrements of an additional compressor and expander, the actual efficiency would fall off.

Perhaps, as has been noted, the applications for the simple Joule cycle will be limited to space stations, where heat can be rejected to outer space without qualification. In fact, the Joule cycle has been studied experimentally for this particular application.

Internal Heat Regeneration. As an alternative, the Joule cycle can be operated with internal heat regeneration, whereby the expanded working fluid is used to heat the compressed working fluid. The effect is the same as for the Brayton cycle: the cycle efficiency increases as the compression–expansion ratio is lowered (as will be shown subsequently).

3.6. REGENERATIVE CYCLE

Corresponding to the open Brayton cycle with heat regeneration, the closed Joule cycle can be adapted similarly as shown in Fig. 3.2. Note that heat transfer occurs after compression. This makes compression more effective than it would be if the temperature was raised before compression; that is, if compression occurred after heat transfer, then the theoretical compression requirement would be

$$T_2 - T_1' = T_1'(\alpha - 1),$$

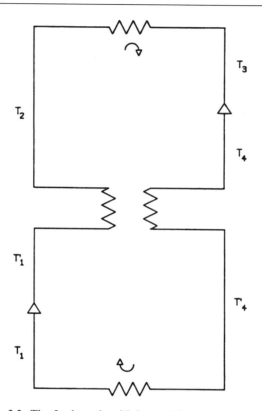

Fig. 3.2. The Joule cycle with internal heat regeneration.

whereas if compression occurs first, the requirement is

$$T_1' - T_1 = T_1(\alpha - 1).$$

Since $T_1' > T_1$, the latter requirement is less than the former.
    In the limit, as $T_2 \to T_4$, the efficiency becomes

$$\text{Eff}_{\text{regen}} = \frac{(T_3 - T_4) - (T_1' - T_1)}{T_3 - T_2}$$

$$= \frac{(T_3 - T_2) - (T_1' - T_1)}{T_3 - T_2}$$

$$= 1 - \frac{T_1(\alpha - 1)}{T_3 - T_2}.$$

This result is identical to that for the Brayton cycle with heat regeneration. Again, the efficiency increases as α decreases.

Rankine Cycle with Internal Heat Regeneration

In the Rankine cycle version of the foregoing regeneration cycle, the compressor would be replaced with a pump and $T_1' \sim T_1 \sim T_4'$. The efficiency would be given by

$$\text{Eff}_{\text{Rankine}} = \frac{C_p(T_3 - T_4)}{\Delta H + C_p(T_3 - T_4)} = \frac{T_3 - T_4}{\Delta H / C_p + (T_3 - T_4)},$$

where $T_4 \geq T_2$. Raising T_4 (and T_2) will cause the efficiency to fall off. Thus internal heat regeneration does not increase the efficiency of the Rankine cycle, but rather lowers it.

In the limit, $T_2 \sim T_4$ and an increase in the temperature level T_4 will still have the same effect of lowering the efficiency.

3.7. EFFICIENCY WITH TWO-STAGE COMPRESSION

The Joule cycle with two-stage compression is diagrammed in Fig. 3.3. Interstage cooling occurs between T_a and T_b.

The theoretical efficiency may be derived as

$$\text{Eff}_{\text{two-stage}} = \frac{(T_3 - T_4) - (T_a - T_1) - (T_2 - T_b)}{T_3 - T_2}$$

$$= \frac{(T_3 - T_4) - (T_2 - T_1) - (T_a - T_b)}{T_3 - T_2}$$

$$= \frac{(T_3 - T_2) - (T_4 - T_1)}{T_3 - T_2} - \frac{(T_a - T_b)}{T_3 - T_2}$$

$$= 1 - \frac{1}{\alpha} - \frac{(T_a - T_b)}{T_3 - T_2}.$$

Thus the efficiency will be lower than for single-stage compression due to the presence of the term $(T_a - T_b)$, where $T_a > T_b$. If $T_a = T_b$, then of course the operation is equivalent to single-stage compression.

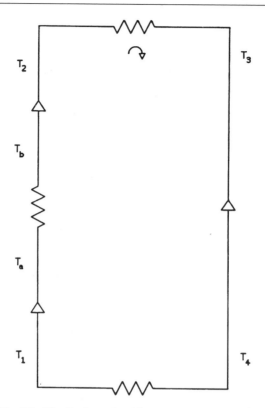

Fig. 3.3. The Joule cycle with two-stage compression.

Observe that

$$\frac{T_a}{T_1} = \left(\frac{P_a}{P_1}\right)^{k/(k-1)} = \alpha_1, \qquad \frac{T_2}{T_b} = \left(\frac{P_2}{P_b}\right)^{k/(k-1)} = \alpha_2$$

and

$$\frac{T_2}{T_1}\frac{T_a}{T_b} = \left(\frac{P_2}{P_1}\right)^{k/(k-1)} = \alpha_1\alpha_2 = \alpha \quad \text{or} \quad \frac{T_2}{T_1} = \alpha\frac{T_b}{T_a}.$$

Furthermore, for two-stage compression,

$$-\overline{w} = C_p(T_a - T_1) + C_p(T_2 - T_b) = C_p(T_2 - T_1) + C_p(T_a - T_b)$$

or

$$-\overline{w} = C_p T_1 \left[\alpha \frac{T_b}{T_a} - 1 \right] + C_p(T_a - T_b) = C_p T_2 \left[1 - \frac{T_a/T_b}{\alpha} \right] + C_p(T_a - T_b)$$

and

$$\overline{q} = -C_p(T_a - T_b),$$

$$\overline{q} - \overline{w} = \Delta H = C_p(T_2 - T_1).$$

Also, on substituting for α,

$$\text{Eff} = 1 - \frac{T_b/T_a}{T_2/T_1} - \frac{T_a - T_b}{T_3 - T_2} = 1 - \frac{T_b}{T_a}\frac{T_1}{T_2} - \frac{T_a - T_b}{T_3 - T_2}.$$

Generally speaking, if T_1 is constant and T_2 remains constant, then the work requirement $(-w)$ will be greater for the cycle with two-stage compression than for single-stage compression. The effect on the efficiency is less obvious, however.

Similarly, if the overall compression ratio remains constant, then α will be constant. The effect on the work requirement will be more circumspect, although it is generally acknowledged that, for the same overall compression ratio, the work requirement will be less for multistage compression with intercooling than for adiabatic single-stage compression without cooling. The efficiency obviously will be less for the cycle with two-stage compression than for single-stage compression. As a consequence, more involved procedures are required in order to compare the one with the other.

For the theoretical efficiency with two-stage compression, alternate expressions may be provided as follows:

$$\text{Eff} = 1 - \frac{C_p(T_4 - T_1) + C_p(T_a - T_b)}{C_p(T_3 - T_2)}$$

$$= 1 - \frac{T_3/\alpha - (T_2/\alpha)(T_a/T_b)}{T_3 - T_2} - \frac{T_a - T_b}{T_3 - T_2}$$

$$= 1 - \frac{1}{\alpha}\frac{T_3 - T_2(T_a/T_b)}{T_3 - T_2} - \frac{T_a - T_b}{T_3 - T_2}.$$

In comparing the preceding expressions with single-stage compression, there are two circumstances of interest: in the one, the final temperatures

T_2 would be the same; in the other, the compression ratios α would be the same. Note that when $T_a = T_b$, the results for two-stage compression merge with single-stage compression. The problem can be viewed as that of holding T_1, T_3, and T_b constant while varying α with T_2 constant, or else varying T_2 with α constant.

The Final Compression Temperature Is the Same

Here, T_2 will be the same constant value for two-stage and single-stage compression. Thus for the effect of varying α or T_a,

$$-dw = C_p \, dT_a.$$

Therefore, the work requirement $(-w)$ will tend to increase as T_a increases, that is, as more interstage cooling is used. For the two-stage cycle efficiency,

$$
\begin{aligned}
d(\text{Eff}) &= \frac{-dT_4 - dT_a}{T_3 - T_2} = \frac{-T_3 d(1/\alpha) - dT_a}{T_3 - T_2} \\
&= \frac{T_3(1/\alpha)(1/T_a) \, dT_a - dT_a}{T_3 - T_2} = \frac{(T_4/T_a - 1)}{T_3 - T_2} \, dT_a.
\end{aligned}
$$

If $T_4 < T_a$, then the efficiency will decrease as T_a increases. The boundary condition is that, when $T_a = T_b$, the two-stage cycle efficiency is equal to the single-stage cycle efficiency.

The Overall Compression Ratio Is the Same

Here, α will be the same, whereby T_2 and T_a will vary. Thus

$$
\begin{aligned}
-d\overline{w} &= C_p \, dT_2 + C_p \, dT_a = C_p T_1 \alpha \frac{-T_b}{T_a^2} \, dT_a + C_p \, dT_a \\
&= C_p \left[-\frac{T_2}{T_a} + 1 \right] dT_a.
\end{aligned}
$$

Since $T_2 > T_a$, then the compression requirement $(-w)$ will decrease as T_a increases (and T_2 decreases).

With regard to the two-stage cycle efficiency, since both T_4 and T_1 will be constants, then

$$d(\text{Eff}) = -\frac{dT_a}{T_3 - T_2} - \frac{T_a - T_b}{(T_3 - T_2)^2} dT_2.$$

Moreover,

$$dT_2 = T_1 \alpha \frac{-T_b}{T_a^2} dT_a = \frac{-T_2}{T_a} dT_a.$$

Therefore,

$$d(\text{Eff}) = \frac{1}{T_3 - T_2} \left[-1 + \frac{T_a - T_b}{T_3 - T_2} \frac{T_2}{T_a} \right] dT_a.$$

Whether or not the efficiency increases or decreases as T_a increases will depend upon the sign of the term in brackets. Initially, for small values of $T_a - T_b$, the efficiency will decrease as T_a increases.

3.8. COMBINED CYCLES

If a second, reverse Joule cycle were to be used as a heat pump, then let its added configuration and notation be as shown in Fig. 3.4. Furthermore, let the temperature ratios be in terms of β, where

$$\beta = \frac{T_2'}{T_1'}, \qquad \beta = \frac{T_3'}{T_4'}.$$

If heat transfer occurs from the interval $T_4 - T_1$ to $T_1' - T_4'$ and from $T_3' - T_2'$ to $T_2 - T_3$, then the combined efficiency is

$$
\begin{aligned}
\text{Eff}_{\text{combined}} \\
&= \frac{(T_3 - T_4) - (T_2 - T_1) - (T_2' - T_1') - (T_3' - T_4')}{(T_3 - T_2) - (T_3' - T_2')} \\
&= \frac{T_3(1 - 1/\beta) - T_2(1 - 1/\beta) + T_2'(1 - 1/\beta) - T_3'(1 - 1/\beta)}{(T_3 - T_2) - (T_3' - T_2')} \\
&= \frac{(T_3 - T_2)(1 - 1/\beta) - (T_3' - T_2')(1 - 1/\beta)}{(T_3 - T_2) - (T_3' - T_2')}.
\end{aligned}
$$

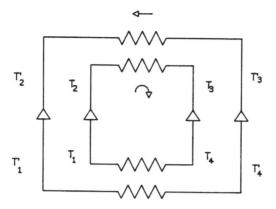

Fig. 3.4. Use of a reverse Joule cycle as a heat pump.

If $\beta = \alpha$, then the theoretical efficiency would be

$$\text{Eff}_{\text{combined}} = 1 - \frac{1}{\beta} = 1 - \frac{1}{\alpha},$$

which is the same as for the Joule cycle alone. If $\beta > \alpha$, which would be required for heat transfer to occur in the correct direction, then the efficiency would be less than for the Joule cycle per se. If $\beta < \alpha$, then the combined cycle would not function as intended.

If the heat from $T_3' - T_2'$ was rejected to the surroundings, then the combined efficiency would be

$$\text{Eff}_{\text{combined}} = \frac{(T_3 - T_4) - (T_2 - T_1) + (T_2' + T_1') - (T_3' - T_4')}{T_3 - T_2}$$

$$= \frac{(T_3 - T_2)(1 - 1/\alpha) - (T_3' - T_2')(1 - 1/\beta)}{T_3 - T_2}$$

$$= \frac{\alpha - 1}{\alpha} - \frac{T_3' - T_2'}{T_3 - T_2} \frac{\beta - 1}{\beta}.$$

Also, if $(T_4 - T_1) \sim (T_4' - T_1')$, then

$$\text{Eff}_{\text{combined}} \sim 1 - \frac{T_3' - T_2'}{T_3 - T_2}.$$

Equating the latter two expressions for efficiency and collecting terms we find that

$$\frac{T_3' - T_2'}{T_3 - T_2} = \frac{\beta}{\alpha},$$

so that

$$\text{Eff}_{\text{combined}} \sim 1 - \frac{\beta}{\alpha}.$$

Thus if $\beta = \alpha$, the combined efficiency becomes zero. If $\beta > \alpha$, then the efficiency is negative. Only if $\beta < \alpha$ is the operation practical. In fact, if

$$\frac{\beta}{\alpha} = \frac{1}{\alpha'},$$

where α' pertains to the case where heat from the primary cycle is rejected to the surroundings, then the combined cycle and single cycle become equivalent.

3.9. WASTE-HEAT RECAPTURE

The Joule cycle would become more efficient if the rejected waste-heat could be put back into the forefront of the cycle. The use of a heat-pump cycle appears to be self-defeating, as previously explained.

As an alternative, the combustion system itself can be used to recapture the waste-heat and return it to the forefront of the cycle. In effect, the combustants, notably the air, are preheated from ambient conditions via indirect heat transfer with the exhaust gases from the turbine-expander. The rejected heat, in sum, is used to preheat the combustants, that is, the fuel–air mixture or the combustive air. In turn, the combustants are ignited and combusted in the combustion zone of a process heater or tube-still heater; the working fluid is on the tube side. The compressed working fluid is thus raised to temperature for expansion. The combustion product gases are returned as nearly as possible to ambient conditions by indirect heat transfer with the incoming compressed working fluid in the convective section of the heater.

In summary, then, the net effect is to recycle the rejected heat from a lower temperature level to a higher temperature level. This rejected waste-heat therefore becomes a supplementary (or complementary) heat input to heat the compressed working fluid. This process is diagrammed in Fig. 3.5.

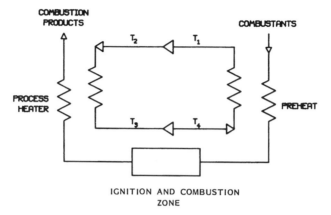

Fig. 3.5. Waste-heat recapture.

Refinements could include the further utilization of the combustion product off-gases or stage gases for a low-grade heat source such as for space heating.

Ideally speaking, the efficiency could be represented by

$$\text{Eff}_{\text{recycle}} = \frac{(T_3 - T_4) - (T_2 - T_1)}{(T_3 - T_2) - (T_4 - T_1)},$$

where the denominator denotes the net heat input to the working fluid. In terms of α,

$$\text{Eff}_{\text{recycle}} = \frac{T_3(1 - 1/\alpha) - T_2(1 - 1/\alpha)}{(T_3 - T_2) - (T_3/\alpha - T_2/\alpha)}$$

$$= \frac{1 - 1/\alpha}{1 - 1/\alpha} = 1.$$

The theoretical efficiency therefore approaches unity or 100%.

In the foregoing derivation it is assumed that the exit temperature of the combustion products approaches T_2. For a representative combustion system, $T_2 \sim 200 + 460 = 660°R$. Under standard conditions (25°C or 77°F), for the standard heat of combustion, $T_0 = 77 + 460 = 537°R$. The difference represents a loss in fuel efficiency, which will affect the overall cycle efficiency, as will the mechanical efficiency losses in compression and expansion.

Two-Phase Cycle

As an alternative, the working fluid can enter the two-phase region; thus the Rankine cycle is employed in lieu of the Joule cycle. Under this circumstance, for a single component, a region of constant temperature is incurred between T_4 and T_1, where condensation occurs. Similarly, a region of constant temperature will exist between T_2 and T_3, where vaporization occurs. Sensible heat changes are, of course, incurred also. The foregoing situation will require that T_1 be considerably above the ambient temperature of the combustants, that is, the fuel–air mixture or the combustive air, for the combustants to absorb this latent heat change. No special complications arise in the process heater.

Most significantly, however, the compressor is replaced by a pump. Thus as a simplification, $T_2 \sim T_1$.

The efficiency will, therefore, become, ideally,

$$\text{Eff}_{\text{recycle}} = \frac{T_3 - T_4}{(T_3 - T_2) - (T_4 - T_1)}$$

$$= \frac{T_3(1 - 1/\alpha)}{(T_3 - T_4) - (T_2 - T_1)}$$

$$= \frac{T_3(1 - 1/\alpha)}{T_3(1 - 1/\alpha)} = 1.$$

Again, the theoretical efficiency is unity for recycling the waste heat via the Rankine cycle. This efficiency, of course, would be reduced by the mechanical and fuel efficiency losses and by the temperature approaches in the waste-heat exchanger and process heater.

To keep T_2 as low as possible, it is advisable that the working fluid have a relatively low heat of vaporization (or condensation). Organic compounds meet this stipulation (e.g., circa 150 Btu/lb versus up to 1000 Btu/lb for the water–steam system). Unfortunately, organics will break down at temperatures higher than about 800°F; therefore, a gas–liquid separator at T_1 is required to remove the gaseous decomposition products, which can be used as a gaseous fuel, even for the process heater.

Excess Air Requirements

The problem exists that large volumes of excess air are required to absorb the waste heat, and any excess air that is heated above ambient conditions will lower the fuel efficiency. This loss will more than offset the gains made in cycle efficiency.

3.10. CONTINUOUS NONADIABATIC BEHAVIOR DURING COMPRESSION AND EXPANSION

An idealized cycle may be constituted similarly to the Joule cycle except that the compression and expansion stages per se are nonadiabatic rather than isentropic and adiabatic. The sequence is illustrated schematically in Fig. 3.6 in *P-T* coordinates. The subject was previously discussed in Section 2.7.

As a hypothetical embodiment, heat would be rejected during the compression step and heat would be added during expansion through a turbine or mechanical expander. Means for heat transfer would, of course, be required, which is a formidable problem from a practical standpoint.

In general, the degree of heat exchange is arbitrary, and any sort of *P-T* behavior may be prescribed during compression and expansion and may be in either direction: rejection and/or addition. Here, however, heat is rejected from the working fluid during compression and added to the working fluid during expansion. As a special or limiting case, the expansion and compression each may be assumed isothermal or one or the other may be isothermal. Again we emphasize that the heat rejection–addition occurs during compression and/or expansion, as distinguished from interstage heat transfer, which will be developed in Chapter 4 as the more practical embodiment.

Referring to Fig. 3.6 in the limit the arbitrary compression behavior along leg 1-2 may merge with the arbitrary expansion behavior along leg 3-4, whereby points 1 and 2 become common, as do points 3 and 4. In other words, all of the heat rejection and addition occurs only during compression and expansion. This, in fact, is a feature of the Ericsson cycle, except in the Ericsson cycle isothermal behavior is specified, so that the

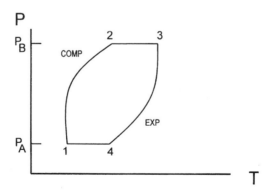

Fig. 3.6. Nonadiabatic Joule cycle.

expander temperature is higher than the compressor temperature and, in principle, heat regeneration can be used.

In the analysis of the cycle, the temperature variable during the compression step will be denoted as T_c, which will vary between the limits T_1 and T_2. Similarly, the pressure during compression will be denoted P_c and will vary between P_A and P_B.

During the expansion step, the temperature variable will be denoted as T_e and will vary between the limits T_3 and T_4. The pressure variable will in turn be denoted as P_e and will vary between P_B and P_A. The foregoing distinctions are necessary because two different domains of variation are involved that relate to two distinct steps or operations.

As previously derived in Section 1.8, during compression, for a perfect gas,

$$-dw_{comp} = (C_p - f)\, dT_c = \frac{RT_c}{P_c}\, dP_c,$$

where $f = f(T_c)$ denotes the arbitrary heat exchange with the surroundings and where the generalized irreversibility $d\omega^*$ is assumed equal to zero. More completely,

$$\frac{dq_{comp}}{dT_c} = f = f(T_c) \quad \text{or} \quad f = f_c = f_c(T),$$

where f is negative for heat rejection in the surroundings, whereas dT and dP are positive. From the energy balance,

$$\int_1^2 \frac{C_p - f}{T_c}\, dT_c = \int_A^B \frac{R}{P_c}\, dP_c$$

and

$$-\overline{w}_{comp} = \int_1^2 (C_p - f)\, dT_c, \qquad q_{comp} = \int_1^2 f\, dT_c.$$

Furthermore,

$$\Delta H = \overline{q} - \overline{w} = C_p(T_2 - T_1),$$

where C_p is assumed constant. The enthalpy of the working fluid will increase regardless of the behavior of f. If $f = f_c$ is a constant, however,

$$\frac{T_2}{T_1} = \left(\frac{P_B}{P_A}\right)^{R/(C_p - f)},$$

which is of the same form as for an isentropic compression and becomes identical if $f_c = 0$.

During an expansion where dT and dP_e are negative,

$$-dw_{\exp} = (C_p - g)\, dT_e = \frac{RT_e}{P_e}\, dP_e,$$

where the arbitrary function $g = g(T_e)$ is given by

$$\frac{dq_{\exp}}{dT_e} = g = g(T_e) \quad \text{or} \quad g = f_e = f_e(T).$$

For heat *added* to the working fluid, g will be negative. In turn,

$$\int_3^4 \frac{C_p - g}{T_e}\, dT_e = \int_B^A \frac{R}{P_e}\, dP_e$$

and

$$-\overline{w}_{\exp} = \int_3^4 (C_p - g)\, dT_e, \qquad \overline{q}_{\exp} = \int_3^4 g\, dT_e.$$

Also

$$\Delta H = \overline{q} - \overline{w} = C_p(T_4 - T_3).$$

The enthalpy of the working fluid will decrease during the expansion.

Again, if $g = f_e$ is constant, the equation form for an isentropic adiabatic expansion will be obtained, and if $g = f_e = 0$, the expansion becomes identically isentropic.

Note that it is required that

$$\int_1^2 \frac{C_p - f}{T_c}\, dT_c = -\int_3^4 \frac{C_p - g}{T_e}\, dT_e = \int_4^3 \frac{C_p - g}{T_e}\, dT_e,$$

which can be reduced further if f or g is constant. Alternately, the preceding equation can be presented as

$$C_p \ln\left(\frac{T_2}{T_1} \frac{T_4}{T_3} \right) = \int_1^2 f\, dT_c + \int_3^4 g\, dT_e.$$

If $T_2/T_1 = T_3/T_4$, then the term on the left-hand side of the preceding equation becomes zero.

For the theoretical cycle efficiency,

$$\text{Eff} = \frac{\overline{w}_{\text{exp}} + \overline{w}_{\text{comp}}}{\overline{q}_{\text{exp}} + C_p(T_3 - T_2)}$$

$$= \frac{-\int_3^4 (C_p - g)\, dT_e - \int_1^2 (C_p - f)\, dT_c}{\int_3^4 g\, dT_e + C_p(T_3 - T_2)}.$$

In the foregoing notation, $w_{\text{net}} = w_{\text{exp}} + w_{\text{comp}}$.

The determination of the efficiency will depend upon what simplifications can be made. For instance, if $T_2/T_1 = T_3/T_4 = \alpha$, as is the case for the ordinary Joule cycle with isentropic adiabatic compression and expansion, then, as previously determined,

$$\int_3^4 g\, dT_e = \int_1^2 f\, dT_c$$

and

$$\text{Eff} = \frac{C_p(T_3 - T_4) - C_p(T_2 - T_1)}{-\int_4^3 g\, dT_e + C_p(T_3 - T_2)} = \frac{(1 - 1/\alpha)C_p(T_3 - T_2)}{-\int_4^3 g\, dT_e + C_p(T_3 - T_2)},$$

where g is negative and $T_3 > T_4$. Furthermore, w_{net} reduces to the expression denoted by the denominator, which is the same as for the conventional Joule cycle. It may be ascertained, also, that if $g = 0$, then the efficiency reduces to that of the convention Joule cycle.

When g is negative (heat is added to the expansion), the cycle efficiency becomes less than for the conventional Joule cycle. In addition, a negative g implies a negative f under the foregoing circumstances, which indicates that heat is rejected during compression.

If f and g both are specified as constants, then we obtain

$$g \ln \left(\frac{T_3}{T_4} \right) = f \ln \left(\frac{T_2}{T_1} \right)$$

from which $f = g$. If the condition that $T_3 = T_2$ and $T_4 = T_1$ is specified, then the efficiency becomes zero. In general, however, neither f nor g has to be constant or equal.

Heat Recapture by the Combustion System

Suppose that the heat rejected during compression could be used to preheat the air or combustants used to supply heat, including the heat added during expansion. This situation would encompass legs 2-3 and 3-4. In this way the rejected heat is captured by the combustants and recycled via the heat of combustion and its further utilization. Depending upon the temperature approach(es), the overall efficiency thereby can be raised, notably via concurrent heat transfer between the combustion system and the working fluid.

By necessity, therefore, concurrent heat transfer should be a requirement in order that the initial combustion reactants and final combustion products tend to approach the same (ambient) temperature level. If the temperature levels are exactly the same, of course, this is the criterion for 100% overall efficiency. Thus the combustion system—the reactants and products—would be required to follow the temperature profile of the working fluid, first up and then down. The difference in temperature between the combustion system and the working fluid, or the approach entering and leaving, is a measure of the degree of departure from 100% overall efficiency. The departure is affected also by compressor and expander losses.

This subject is explored in detail in Chapter 4 using the concept of the isentropic adiabatic Joule cycle for each stage, but with interstage cooling between compressor stages and with interstage heating between expander stages.

3.11. HEAT BALANCES

The determination of the overall efficiency will require the establishment of the fuel efficiency. The fuel efficiency will require heat balances between the cycle proper and the combustion system used. Furthermore, it is necessary to introduce the mechanical efficiency, for example, based on the performance of the compressor and expander. Pressure-drop losses in the heat exchangers and lines may be included in the mechanical efficiency. The fuel efficiency or the mechanical efficiency, or both, may become controlling, depending upon the particular embodiment.

Heats of Reaction (Combustion)

The combustion temperatures reached depend upon the heat of combustion and the temperature of the reactants or combustants.

The enthalpy difference, at constant pressure, between the reactants R and products P of a chemical conversion, per mole of reactant or combustant, is

$$H_P - H_R = \Delta H_0 + \Sigma P(C_p)_P (T_P - T_0) - \Sigma R(C_p)_R (T_R - T_0),$$

where 0 denotes the standard reference condition (25°C or 77°F) and ΔH_0 is the standard heat of reaction or combustion. The difference, $\Delta H = H_P - H_R$, may be denoted as the heat of reaction at the stated conditions. Most usually, pressure adjustments are not a consideration, and the preceding correction is for temperature effects only.

It will be understood for the purposes here that the heat of reaction is the net heat of combustion rather than the gross heat. In other words, any water produced as a reaction product will remain in the gaseous phase, such that the latent heat of condensation of any water vapor produced is not incorporated to yield the gross heat of combustion.

It may be noted also, that the heat delivered during combustion will be affected by the water or moisture content of the fuel. In other words, part of the heat of combustion is required to vaporize the moisture and superheat it to reaction conditions. This is partially offset by the fact that the presence of water or moisture enhances fuel reactivity or combustibility. There is a steam-reforming action between the fuel and the water vapor which yields reactive hydrogen and carbon monoxide as the actual combustibles.

If $\Sigma P \sim \Sigma R$ and $(C_p)_P \sim (C_p)_R$, or if

$$\Sigma P(C_p)_P \sim \Sigma R(C_p)_R,$$

then

$$H_P - H_R = \Delta H_0 + \Sigma P(C_p)_P (T_P - T_R),$$

and for an adiabatic reaction,

$$0 = \Delta H_0 + \Sigma P(C_p)_P (T_P - T_R).$$

If the reaction is exothermic, ΔH will have a negative value.

Combustion Stoichiometry

The combustion stoichiometry, in moles, can be represented as

$$\Sigma R \rightarrow \Sigma P \quad \text{or} \quad W_R \rightarrow W_P$$

or

$$W_F + W_{O_2} + W_{N_2} \rightarrow W_P,$$

where F denotes the fuel, O_2 denotes oxygen, and N_2 denotes the nitrogen in the combustive air. The oxygen requirement will be stipulated as n moles of oxygen per mole of fuel; that is,

$$W_{O_2} = nW_F$$

and where, approximately, for the nitrogen and air,

$$W_{N_2} = 4nW_F,$$

$$W_{air} = 5nW_F.$$

The foregoing expressions are the theoretical requirements, excluding excess air.

Strictly speaking, the value of W_P or ΣP will not necessarily be the same as for the reactants. A few comparisons are shown in Table 3.2. For most practical purposes, however, this approximation is good enough. Note that for the combustion of solid fuels, such as carbonaceous solids, the fuel may or may not be included in the (gaseous) reactants. Also, a fuel composition designatable as $(C_2H_2)_x$ will approximate the makeup of liquid hydrocarbon fuels.

Excess Air

Any excess air used may be designated by the difference between the stack gases produced and the totality of the products of combustion—the carbon dioxide, water vapor, and the nitrogen of the stoichiometric air requirement. This difference may be stated as $W_S - W_P$, where W_S would be the molar flow rate of the stack gases and W_P is the totality of the products of combustion, including the nitrogen of the stoichiometric air requirement.

Alternatively, the excess air may be expressed as the difference between the total molar fuel–air mixture, including the excess air, and the stoichiometric fuel–air mixture; that is, the difference is $W_A - (5n)W_F - W_F$, where W_A is the total molar flow rate for the fuel–air mixture or combustants used, including the excess air, and W_F is the molar flow rate of the fuel. The stoichiometric air requirement is $5n(W_F)$ and the stoichiometric fuel–air mixture is $(5n)W_F + W_F = (5n + 1)W_F$.

Table 3.2. COMPARISON OF COMBUSTION STOICHIOMETRIES

$$C + O_2 + (4N_2) \rightarrow CO_2 + (4N_2)$$
$$CH_4 + 2O_2 + (8N_2) \rightarrow CO_2 + 2H_2O(g) + (8N_2)$$
$$H_2 + \tfrac{1}{2}O_2 + (2N_2) \rightarrow H_2O(g) + (2N_2)$$
$$CO + \tfrac{1}{2}O_2 + (2N_2) \rightarrow CO_2 + (2N_2)$$
$$C_2H_2 + \tfrac{5}{2}O_2 + (10N_2) \rightarrow 2CO_2 + H_2O(g) + (10N_2)$$

Fuel	W_F	n	ΣR	ΣP	$\Sigma R/\Sigma P$
C	1	1	5 *[a]	5	1
CH_4	1	2	11	11	1
H_2	1	1/2	2.5	3	1.2
CO	1	1/2	2.5	3	1.2
C_2H_2	1	5/2	12.5	13	1.04

[a]Pertains only to gaseous reactants

From the foregoing information, the excess air may be expressed as

$$W_S - W_F = W_A - (5n + 1)W_F.$$

If, however, it is assumed that $W_S \sim W_A$, then the molar flow rate W_P of the combustion products is the same as for the stoichiometric fuel–air mixture:

$$W_P \sim (5n + 1)W_F.$$

This simplification will be assumed to hold, for most practical purposes. Any excess air used, if it is of interest, may be restated as a percentage of the stoichiometric air requirement $(5n)W_F$. (Observe that if mass flow rates were used for W instead of molar flow rates, then W_S would be exactly equal to W_A.)

Combustion Heat Balance

The heat balance for a combustion system may be written as

$$0 = W_F \, \Delta H_0 + W_S(C_p)_S(T_c - T_b), \qquad (1)$$

where here

W_F = moles per hour of fuel,

W_S = moles per hour of combustion products or stack gases ($W_S = W_P = \Sigma P$),

T_c = adiabatic combustion temperature reached,

T_b = temperature of the combustants.

If the combustants are at ambient temperature, then $T_b = T_a$, where T_a is the ambient temperature. The total moles of combustion products are assumed to be equal to the total moles of combustants, including excess air, and the heat capacities are assumed to be equivalent; that is,

$$W_S = W_A \quad \text{and} \quad \Sigma P = \Sigma R$$

or

$$W_S = W_F \Sigma P \quad \text{and} \quad W_A = W_F \Sigma R,$$

where W_A is the moles per hour of the total reactants or combustants, including any excess air. Also, the moles of reactants R and products P are per mole of fuel. Furthermore, $W_S(C_p)_S = W_A(C_p)_A$.

An additional equation also may be written:

$$W_S(C_p)_S(T_c - T_S) = W_{WF}(C_p)_{WF}(T_3 - T_2), \tag{2a}$$

where

$$W_{WF} = \text{moles per hour of working fluid circulating,}$$

$$(C_p)_{WF} = \text{molar heat capacity of working fluid,}$$

$$T_S = \text{stack-gas temperature.}$$

It is generally assumed that, for maximum heat transfer, the combustants move countercurrently with respect to the working fluid.

Rankine Cycle. If the Rankine cycle is utilized, the additional equation will appear as

$$W_S(C_p)_S(T_c - T_S) = W_{WF}\left[\Delta H + \left(\overline{C}_p\right)_{WF}(T_3 - T_2)\right], \tag{2b}$$

where sensible and latent heat changes for the working fluid are incorporated into the term in brackets, and $(\overline{C}_p)_{WF}$ is the mean heat capacity.

Heat Balance for Process Heater or Boiler

The heat balance at the process heater, utilizing the Joule cycle, can be stated as

$$0 = W_F \Delta H_0 + W_{WF}(C_p)_{WF}(T_3 - T_2) + W_S(C_p)_S(T_S - T_b). \tag{3a}$$

If ambient air is used for the combustants, then $T_b = T_a$, the ambient air temperature.

Substituting from Equations 1 and 2a,

$$1 = \frac{T_c - T_S}{T_c - T_b} + \frac{T_S - T_b}{T_c - T_b}. \tag{4}$$

The two terms on the right-hand side can be construed as the proportions of the heat of combustion that go to the working fluid and to the stack gases.

Rankine Cycle. For comparison, the heat balance at the boiler, using the Rankine cycle, can be stated similarly:

$$0 = W_F \, \Delta H_0 + W_{WF}\left[\Delta H + \left(\overline{C}_p\right)_{WF}(T_3 - T_2)\right] + W_S(C_p)_S(T_S - T_b), \tag{3b}$$

where sensible and latent heat changes are incorporated into the term in brackets.

Substituting from Equations 1 and 2b, it again follows that

$$1 = \frac{T_c - T_S}{T_c - T_b} + \frac{T_S - T_b}{T_c - T_b}, \tag{4}$$

the same result as obtained for the Joule cycle.

In effect, in either case, there are two independent equations. The other two equations will be dependent.

Fuel Efficiency

The fuel efficiency can be designated by

$$\text{Eff}_{fuel} = \frac{T_c - T_S}{T_c - T_b}. \tag{5}$$

This applies to both the Joule and Rankine cycles. If $T_s \rightarrow T_b$, then hypothetically 100% fuel efficiency could be attained. This is the argument for using heat regeneration.

As an example, employing Equation 5, if

$$T_S = 500 + 460 = 960°R,$$

$$T_c = 3000 + 460 = 3460°R,$$

$$T_b = 100 + 460 = 560°R,$$

then the fuel efficiency would be 86.2%. If T_S were 300°F, then the fuel efficiency would be raised to 93%. Adjustments can be made for heat losses to the surroundings.

Mechanical Efficiency

For the Joule cycle, the effect of compressor and expander mechanical efficiency losses can be incorporated as follows:

$$\text{Eff}_{\text{mech}} = \frac{E_{\text{exp}}(T_3 - T_4) - (1/E_{\text{comp}})(T_2 - T_1)}{(T_3 - T_4) - (T_2 - T_1)},$$

where

$$E_{\text{exp}} = \text{fractional efficiency of turbine-expander},$$

$$E_{\text{comp}} = \text{fractional efficiency of compressor}.$$

On introducing α,

$$\text{Eff}_{\text{mech}} = \frac{E_{\text{exp}}T_3(1 - 1/\alpha) - (1/E_{\text{comp}})T_2(1 - 1/\alpha)}{T_3(1 - 1/\alpha) - T_2(1 - 1/\alpha)}$$

$$= \frac{E_{\text{exp}}T_3 - (1/E_{\text{comp}})T_2}{T_3 - T_2}. \tag{6}$$

If the net mechanical energy is further converted into electrical energy, then the mechanical–electrical efficiency losses can be included in E_{exp}. The mechanical efficiency can be further modified, as per the Joule cycle, by using a mechanically coupled or hydraulically coupled expander and compressor, for example, what may be called a turbo-expander–turbo-compressor.

Rankine Cycle. Here, only expansion losses will be considered; pumping losses are neglected. Thus $E_{\text{mech}} = E_{\text{exp}}$.

Overall Efficiency

The overall efficiency would be

$$\text{Eff}_{\text{overall}} = \text{Eff}_{\text{Joule}} \cdot \text{Eff}_{\text{fuel}} \cdot \text{Eff}_{\text{mech}} \tag{7}$$

and similarly for the Rankine cycle.

Heat Regeneration

If heat regeneration is used between the stack gases or combustion product off-gases and the ambient combustive air or ambient fuel–air mixture, then let

$$W_S(C_p)_S(T_S - T_{Sf}) = W_A(C_p)_A(T_b - T_a),$$

where

T_S = stack-gas temperature before heat regeneration,

T_{Sf} = final stack-gas temperature after heat regeneration,

T_a = ambient temperature,

T_b = temperature of preheated air or preheated fuel–air mixture.

Furthermore,

$$T_S > T_b, \qquad T_{Sf} > T_a.$$

Flow in the heat exchanger or regenerator would ordinarily be countercurrent. For most practical purposes, $W_S(C_p)_S \sim W_A(C_p)_A$. Therefore

$$T_S - T_{Sf} \sim T_b - T_a,$$

or

$$T_S \sim T_{Sf} + (T_b - T_a).$$

The effect on the fuel efficiency is

$$\text{Eff}_{\text{fuel}} = \frac{T_c - T_S}{T_c - T_b} = \frac{T_c - [T_{Sf} + T_b - T_a]}{T_c - T_b}$$

$$= \frac{(T_c - T_b) - (T_{Sf} - T_a)}{T_c - T_b}$$

$$= 1 - \frac{T_{Sf} - T_a}{T_c - T_b}.$$

If $T_{Sf} \to T_a$, then the fuel efficiency would become 100%.
As an example, for a 50°F approach at the regenerator, if

$$T_S = 500°F, \qquad T_b = 450°F,$$

$$T_{Sf} = 150°F, \qquad T_a = 100°F,$$

and if $T_c \sim 3000°F$, then

$$\text{Eff}_{\text{fuel}} = 1 - \frac{150 - 100}{3000 - 450} = 98\%.$$

The problem lies in the size of the heat exchanger required. For a 100°F approach on either end, the fuel efficiency would be about 96%, but an exchanger of less than one-half the size would be required, because the heat load also would be lowered, and so forth. (Strictly speaking, as T_b is reduced, then T_c also will be reduced.)

3.12. COMMENTARY

The Joule cycle, also called the (closed) Brayton cycle (although the latter name also is reserved for gas-fired turbines), has been commented on to a limited degree in the literature. Reference also is found under gas or air engines, or hot air engines, and under air standard cycles. For instance, Reeve (6) (published in 1903) made this observation: "It is the personal opinion of the author that the large gas-engines of the future will operate on no other than the Joule cycle." His reasons, however, are not solely nor even largely based upon the thermodynamic points of view that are developed herein. They are instead based chiefly upon purely mechanical considerations that lie outside the limits of the present discussion. A further comment is worthy of note, however:

[T]he purely thermodynamic aspect of the question shows that under any chosen limits of temperature and pressure, the theoretical efficiency of the Joule cycle ... is greater than that of any other cycles known to have been proposed or put into practice except the Carnot, the Stirling, and the Ericsson. Of these three, all are known by long trial to be impractical.

Doolittle and Zerban (7) (published in 1948) comment as follows:

The mean effective pressure of the Brayton cycle (Joule cycle), although much larger than that of the Carnot cycle, is very low in comparison with that obtained in the Otto, Diesel, and Dual cycles. When reciprocating machinery is used, this low mean effective pressure is a very serious problem. This fact, plus the requirement of two pieces of machinery in the Brayton cycle rather than one, prevented extensive use of the Brayton cycle.

Rotating machines, such as gas turbines, can handle very large volumes per unit of time; hence, for them low mean effective pressures are not a serious handicap. In very recent years, great improvements in the elements of the gas turbine have made them a competitor in the power generation field. The elementary gas turbine follows the Brayton cycle quite closely.

The authors comment further that the reversed Brayton cycle (or Joule cycle) is the basis of air refrigeration systems, that is, of noncondensing working fluid systems.

Ebaugh (8) (published in 1952) made the following statement:

For approximately 50 years after Joule's proposal, the compressor required more power than the turbine developed and hence the cycle was impractical. Modern compressors and turbines range in efficiency up to about 86 percent as compared with reversible adiabatic processes, and these efficiencies make the cycle a practical possibility.

Doolittle (9) again, in another volume published in 1965, further states:

For many years, the so-called hot-air engines were used extensively. The cycle of operation of these engines approximated either the Stirling or Ericsson cycles. These cycles, inherently, have high thermal efficiencies, but much more difficulty was encountered in making the cycle of operation of the actual engine to come at all close to the theoretical. In particular, because of the necessity of transferring heat from air to gas through walls, the large drop in temperature caused a serious loss in efficiency. In addition, the hot-air engines were very large and heavy. Because of these reasons, the use of hot-air engines has been discontinued. Rather recently, however, consideration has been given to the use of these cycles for engines to develop auxiliary power for spacecraft requirements. Some of the difficulties with the hot-air engines may be minimized by the use of very high pressures, possibly 1500 psi or higher.

(*Note*: In the air-standard Stirling cycle, heat is supplied during isothermal expansion, and the air is cooled in a regenerator at constant volume. Heat is rejected during an isothermal compression and the air is reheated in the regenerator at constant volume. The theoretical efficiency is the same as for the Carnot cycle. In the Ericsson cycle, the regenerator is operated at constant pressure.)

In further comment, although the gas-fired turbine, or Brayton cycle, may have advantages for clean fuels, the Joule cycle using indirect heat transfer to the hot working fluid has the advantage when using dirty or low-grade fuels; for example, erosion of the turbine blades is avoided.

Finally, in testing small-scale equipment at bench-scale or pilot plant levels, the turbine expander and compressor likely may possess *lower* efficiencies than large-scale equipment. This lower efficiency must be taken into consideration in evaluating the results.

In calculations using regenerative heat transfer, which presumably would be more applicable to smaller skid-mounted installations, relatively high efficiencies also can be obtained for any of several working fluids investigated, including air. Interestingly, the theoretical efficiency increases as the pressure ratio decreases, within practical limits.

Speaking of regeneration cycles, the waste-heat in the combustion off-gases used as the heat source for the closed-cycle working fluid can be used in an air preheater to reduce efficiency losses, as previously indicated.

REFERENCES

1. Marter, D. H., *Thermodynamics and the Heat Engine*, Thames and Hudson, London, 1960.
2. Wrangham, D. H., *The Theory and Practice of Heat Engines*, Macmillan, New York, 1942.
3. Hougen, O. A., K. M. Watson, and R. A. Ragatz, *Chemical Process Principles. II. Thermodynamics*, 2nd ed., Wiley, New York, 1959.
4. Perry, J. H., Ed., *Chemical Engineers' Handbook*, 4th ed., McGraw-Hill, New York, 1963.
5. *Engineering Data Book*, Gas Processors Suppliers Association, Tulsa OK, 1972.
6. Reeve, S. A., *The Thermodynamics of Heat Engines*, Macmillan, New York, 1903.
7. Doolittle, J. S. and A. M. Zerban, *Engineering Thermodynamics*, International Textbook, Scranton PA, 1948.
8. Ebaugh, N. C., *Engineering Thermodynamics*, 2nd ed., Van Nostrand, New York, 1952.
9. Doolittle, J. S., *Thermodynamics for Engineers*, 2nd ed., International Textbook, Scranton PA, 1964.

CHAPTER 4

The Joule Cycle with Nonadiabatic Compression and Expansion

It is relatively straightforward to derive and demonstrate that there is a net gain of work done by a modified Joule cycle with nonadiabatic differential compression and expansion, as shown in Chapters 2 and 3; that is, the heat removed during compression is more than offset by the heat added during expansion. This concept has been indicated both graphically and by an integration of the differential equations involved. Coupled with a circulating heat transfer medium to which heat is added directly from a combustive source or from an external thermal or geothermal source, the overall efficiency can look very attractive indeed. As previously, explained, the heat transfer medium picks up the heat of compression from the working fluid, to which external heat is then added, and, in turn, heat is transferred back to the working fluid during expansion.

There is, however, an equipment problem with differential or continuous nonadiabatic behavior. A way must be found to remove heat from the working fluid during compression and to add heat during expansion. Conceivably the compressor and expander could be jacketed; that is, a jacket or shroud that has a circulating medium flowing through it, and which is put over the compressor and over the expander, could serve to remove or contribute heat, whichever the case may be.

A much more practical embodiment is to use interstage heat transfer. The equipment items already exist in the form of interstage heat exchangers. Most generally used for interstage cooling between compressor stages, heat exchanger equipment can be adapted as well to interstage heating between expander stages.

Furthermore, it is well known that interstage cooling is beneficial during compression because less work is required for a given pressure ratio. It is less well known that for interstage heating during expansion, more work is produced for a given pressure ratio. Additionally, the pressure ratios for compression and expansion ideally will be identical for the cycle under consideration; that is, for most practical purposes, the high-pressure sides for compression and expansion will have a common value for the pressure, as will the low-pressure sides.

The derivations and calculations for multistage behavior with interstage heat transfer are much more difficult than for the integration of continuous behavior. However, each compressor stage and each expansion stage obeys the relationships for isentropic change, which makes the solution manageable. The problem next may be approached in three ways: by assuming constant inlet or outlet temperatures for compression and for expansion, by assuming constant interstage heat transfer rates during compression and expansion, and by assuming that the interstage heat transfer rates vary geometrically during compression and expansion. (These constant temperatures and rates each would, of course, be different for the compression and expansion sections, with corresponding higher temperatures and rates for the expansion section.) The third-mentioned approach provides the necessary simplification, albeit the solution is sill complicated.

In any event, the final solution that is obtained demonstrates that high conversion efficiencies can be achieved. In fact, in principle and in the limit, it can be demonstrated that all the heat hypothetically can be converted to work in this manner; that is, if the temperature of the heat transfer medium from the expansion section could be made equal to the inlet temperature of the heat transfer medium to the compression section. Such matters require conditions such as heat transfer with no temperature difference and so forth. However, with reasonable temperature differences, remarkably high overall efficiency values can still be achieved.

Both combustive support and thermal support are incorporated into the determinations. Of particular interest is geothermal energy, or any thermal source—even solar energy—to heat the circulating heat transfer medium. The methodology can be adapted as well to nuclear energy sources via generation of superheated steam or superheated pressurized water, which becomes the medium. Operations are adaptable to utilizing process waste-heat as the circulating heat transfer medium or to heat the circulating medium. As a special case, the technology can utilize the waste-heat from the cooler–condensers in conventional steam-power plants using the Rankine cycle. The steam–condensate itself can become the circulating heat transfer medium or, less efficiently, condenser cooling water can be used as the medium. As such, the operation could be referred to as a "bottoming" cycle at the aft end of the Rankine cycle, as distinguished from a "topping" cycle at the fore end. In combined operations there will be an optimum temperature for the waste heat leaving the Rankine cycle and entering the bottoming cycle. This will depend upon the operating conditions vis à vis the relative energy generating capacities and efficiencies for the two cycles; that is, there will be an optimum total energy generating capacity and overall conversion efficiency.

Furthermore, in some ways it is preferable that the circulating heat transfer medium be a liquid, such as a thermal or geothermal fluid, because the working fluid is a gas, and gas–liquid heat transfer occurs at higher rates than gas–gas heat transfer, as would be the case when the heat transfer medium consists of combustive air or air–fuel mixtures and gaseous combustion products.

As a final comment, there is the obvious conclusion that enhanced efficiencies for the conversion of heat to work to electricity translate to conserving and stretching-out our energy resources.

4.1. NONADIABATIC MULTISTAGE COMPRESSION AND EXPANSION

Three scenarios subsequently will be considered, based on interstage cooling and heating:

- The compressor or expander inlet temperature is constant.
- The interstage cooling or heating rate is constant.
- The interstage cooling or heating rate varies geometrically.

A schematic applicable to these scenarios is shown in Fig. 4.1.

The general relationships involved for both compression and expansion are discussed in the subsequent sections.

Adiabatic Compression or Expansion Steps

For each stage in a multistage compression or expansion, the actual compression or expansion step per se will be regarded as adiabatic. The terms stage and step are interchangeable, however, and the convention will be used that "stage" can refer to either or both the actual act of compression and/or the actual act of expansion, or can be used to denote the change from inlet to outlet or from outlet to inlet. For instance, "stage" could denote a series of single-stage compressions or expansions, where the series remains adiabatic. On the other hand, a series of adiabatic compressions or expansions with interstage heat transfer can be referred to as a *section*, a usage to follow herein.

Therefore, referring to Fig. 4.1, for each compressor stage n, let

$$\frac{T_n}{T'_{n-1}} = \left(\frac{P_n}{P_{n-1}} \right)^{(k-1)/k} = \alpha_n$$

or

$$T_n = \alpha_n T'_{n-1}$$

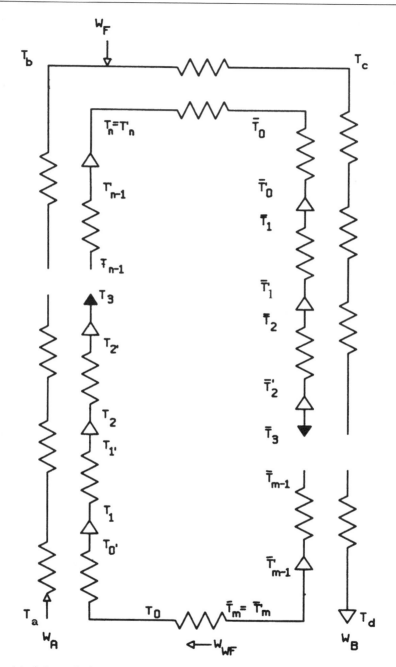

Fig. 4.1. Schematic for multistage compression and expansion with heat recovery using a combustive heat source.

and

$$T_n - T'_{n-1} = (\alpha_n - 1)T'_{n-1} = \left(1 - \frac{1}{\alpha_n}\right)T_n,$$

where the subscript n denotes conditions leaving stage n and $n-1$ denotes conditions entering the stage or step n. Furthermore, the designator T represents the temperature before interstage heat transfer and the designator T' represents the temperature after interstage heat transfer.

If the system is adiabatic, with no interstage heat transfer, then, for instance,

$$T'_{n-1} = T_{n-1}$$

and similarly for the other stages.

For n stages with equal compression ratios,

$$\alpha_1 = \alpha_2 = \alpha_3 = \cdots = \alpha_n$$

and it follows that

$$\left(\frac{P_n}{P_0}\right)^{(k-1)/k} = \left[\left(\frac{P_n}{P_{n-1}}\right)\left(\frac{P_{n-1}}{P_{n-2}}\right)\cdots\left(\frac{P_2}{P_1}\right)\left(\frac{P_1}{P_0}\right)\right]^{(k-1)/k}$$

or

$$\alpha = \alpha_n \alpha_{n-1} \cdots \alpha_2 \alpha_1$$
$$= (\alpha_n)^n,$$

where α pertains to the overall compression ratio P_n/P_0. Alternately,

$$\alpha_n = \alpha^{1/n}.$$

If, for example, the requisite heat of compression is removed between the stages or steps, then the inlet temperatures can be assigned the same constant value. Let this value be T_0. Therefore, $T'_{n-1} = T_0$ and

$$T_n - T'_{n-1} = (\alpha_n - 1)T_0.$$

Furthermore, adding all the temperature differences between T_1 and T_n gives

$$T_n - T_0 = n(\alpha_n - 1)T_0.$$

The preceding difference, when multiplied by C_p, will give the total heat of compression. Note that the heat of compression for each stage remains constant.

Table 4.1. GENERALIZED WORK EXCHANGE IN MULTISTAGE COMPRESSION

$$-W_1 = C_p(T_1 - T_0') = C_pT_1(1 - 1/\alpha_1) = C_pT_0'(\alpha_1 - 1)$$
$$-W_2 = C_p(T_2 - T_1') = C_pT_2(1 - 1/\alpha_2) = C_pT_1'(\alpha_2 - 1)$$
$$-W_3 = C_p(T_3 - T_2') = C_pT_3(1 - 1/\alpha_3) = C_pT_2'(\alpha_3 - 1)$$
$$\vdots$$
$$-W_n = C_p(T_n - T_{n-1}') = C_pT_n(1 - 1/\alpha_n) = C_pT_{n-1}'(\alpha_n - 1)$$

If only a single stage is involved, where $n = 1$, then

$$T_1 - T_0 = (\alpha - 1)T_0,$$

where it is implied that $T_0' = T_0$. When multiplied by C_p, the result would be the heat of compression. The same sort of relationships, in reverse, apply for expansion. A more general accounting follows.

Nonadiabatic Multistage Compression

The generalized stagewise expressions for the work exchange in nonadiabatic multistage compression are presented in Table 4.1. It is understood that $-W$ represents the work done on the working fluid in heat units; that is, $-W$ will be positive. Furthermore, $T_n > T_{n-1}'$. The subscript n denotes the nth stage and α_n pertains to the compression ratio for each stage.

If the α_n are equal, it will follow that

$$-\Sigma W_n = -\sum_1^n W_i = C_p\left(1 - \frac{1}{\alpha_n}\right)\sum_1^n T_i = C_p(\alpha_n - 1)\sum_0^{n-1} T_i'.$$

The generalized stagewise expressions for the heat exchange in nonadiabatic compression are presented in Table 4.2. Here, it is understood that Q represents the heat added to the working fluid in heat units. Therefore, Q will have a negative value for heat removed.

Table 4.2. GENERALIZED HEAT EXCHANGE IN MULTISTAGE COMPRESSION

$$Q_0 = C_p(T_0' - T_0) = C_p(T_1/\alpha_1 - T_0)$$
$$Q_1 = C_p(T_1' - T_1) = C_p(T_2/\alpha_2 - T_1) = C_p(T_1 - \alpha_1 T_0')$$
$$Q_2 = C_p(T_2' - T_2) = C_p(T_3/\alpha_3 - T_2) = C_p(T_2 - \alpha_2 T_1')$$
$$\vdots$$
$$Q_{n-1} = C_p(T_{n-1}' - T_{n-1}) = C_p(T_n/\alpha_n - T_{n-1}) = C_p(T_{n-1} - \alpha_{n-1}T_{n-2}')$$
$$Q_n = C_p(T_n' - T_n) = 0$$

If the α_n are equal, it follows that

$$\Sigma Q_n = \Sigma Q_i = C_p \left[\left(\frac{1}{\alpha_n} \right) \sum_1^n T_i - \sum_0^{n-1} T_i \right] = C_p \left[\sum_1^{n-1} T_i' - \alpha_n \sum_0^{n-2} T_i' \right].$$

Across each stage n, the net enthalpy change of the working fluid is given by

$$Q_{n-1} - W_n = C_p(T_{n-1}' - T_{n-1}) + C_p(T_n - T_{n-1}')$$

$$= C_p(T_n - T_{n-1}).$$

The total enthalpic change of the working fluid will be

$$\sum_0^{1-n} Q_i - \sum_1^n W_i = C_p(T_n - T_0)$$

or, for notational simplicity,

$$\Sigma Q_n - \Sigma W_n = C_p(T_n - T_0).$$

The preceding result may be verified by adding the expressions previously obtained for $-\Sigma W_n$ and for ΣQ_n.

The foregoing expression represents the total enthalpy change of the working fluid. If $T_n > T_{n-1} > T_0$, there is an increase in the enthalpy of the working fluid in the direction of flow. It may be said that more energy is added to the working fluid (via compression) than is rejected by interstage cooling.

Nonadiabatic Multistage Expansion

The same notation is used as for compression, but with bars to denote expansion. Furthermore, the symbol β is used to denote the expansion ratio, where $\beta > 1$, and m denotes the stage.

The generalized expressions for the work exchange in nonadiabatic multistage expansion are presented in Table 4.3. In this case, W represents the work done by the working fluid undergoing expansion and it will have a positive value. Also, $T_m' < T_{m-1}$.

If the β_m are equal, it will follow that

$$-\Sigma \overline{W}_m = \sum_1^m \overline{W}_j = C_p(1 - \beta_m) \sum_1^m \overline{T}_j = C_p \left(\frac{1}{\beta_m} - 1 \right) \sum_0^{m-1} \overline{T}_j'.$$

Table 4.3. GENERALIZED WORK EXCHANGE IN MULTISTAGE EXPANSION

$$-\overline{W}_1 = C_p(\overline{T}_1 - \overline{T}_0') = C_p\overline{T}_1(1 - \beta_1) = C_p\overline{T}_0'(1/\beta_1 - 1)$$

$$-\overline{W}_2 = C_p(\overline{T}_2 - \overline{T}_1') = C_p\overline{T}_2(1 - \beta_2) = C_p\overline{T}_1'(1/\beta_2 - 1)$$

$$-\overline{W}_3 = C_p(\overline{T}_3 - \overline{T}_2') = C_p\overline{T}_3(1 - \beta_3) = C_p\overline{T}_2'(1/\beta_3 - 1)$$

$$\vdots$$

$$-\overline{W}_m = C_p(\overline{T}_m - \overline{T}_{m-1}') = C_p\overline{T}_m(1 - \beta_m) = C_p\overline{T}_{m-1}'(1/\beta_m - 1)$$

The generalized expressions for heat exchange in nonadiabatic expansion are presented in Table 4.4. Because here $\overline{Q}$ denotes the heat added to the working fluid, it will have a positive value.

If the β_m are equal, then

$$\Sigma\overline{Q}_m = \sum_1^m \overline{Q}_j = C_p\left[\beta_m \sum_1^m \overline{T}_j - \sum_0^{m-1} \overline{T}_j\right] = C_p\left[\sum_1^{m-1} \overline{T}_j' - \left(\frac{1}{\beta_m}\right)^{m-2}\sum_0^{m-2} \overline{T}_j'\right].$$

Across each stage m in the direction of flow, the net enthalpy change of the working fluid is

$$\overline{Q}_{m-1} - \overline{W}_m = C_p\left(\overline{T}_m - \overline{T}_{m-1}\right).$$

The total enthalpic change becomes

$$\sum_0^{m-1} \overline{Q}_j - \sum_1^m \overline{W}_j = C_p\left(\overline{T}_m - \overline{T}_0\right)$$

or simply

$$\Sigma\overline{Q}_m - \Sigma\overline{W}_m = C_p\left(\overline{T}_m - \overline{T}_0\right).$$

The preceding result can be confirmed by adding the expressions previously derived for $-\Sigma\overline{W}_m$ and for $\Sigma\overline{Q}_m$.

Table 4.4. GENERALIZED HEAT EXCHANGE IN MULTISTAGE EXPANSION

$$\overline{Q}_0 = C_p(\overline{T}_0' - \overline{T}_0) = C_p(\beta_1\overline{T}_1 - \overline{T}_0)$$

$$\overline{Q}_1 = C_p(\overline{T}_1' - \overline{T}_1) = C_p(\beta_2\overline{T}_2 - \overline{T}_1) = C_p(\overline{T}_1' - \overline{T}_0'/\beta_1)$$

$$\overline{Q}_2 = C_p(\overline{T}_2' - \overline{T}_2) = C_p(\beta_3\overline{T}_3 - \overline{T}_2) = C_p(\overline{T}_2' - \overline{T}_1'/\beta_2)$$

$$\vdots$$

$$\overline{Q}_{m-1} = C_p(\overline{T}_{m-1}' - \overline{T}_{m-1}) = C_p(\beta_m\overline{T}_m - \overline{T}_{m-1}) = C_p(\overline{T}_{m-1}' - \overline{T}_{m-2}'/\beta_{m-1})$$

$$\overline{Q}_m = C_p(\overline{T}_m' - \overline{T}_m) = 0$$

The foregoing expression represents the total enthalpy change of the working fluid undergoing expansion. If $\overline{T}_m < \overline{T}_{m-1} < \overline{T}_0$, there is a decrease in the enthalpy of the working fluid; that is, more energy is lost from the working fluid (via expansion) than is added by interstage heating.

4.2. NONADIABATIC MULTISTAGE BEHAVIOR WITH CONSTANT INLET TEMPERATURES

The relationships derived in Section 4.1 simplify as follows for a constant, uniform inlet temperature, for both compression and expansion.

Compression

If a constant inlet temperature T_0 is assigned, whereby

$$T_0 = T_0' = T_1' = T_2' = \cdots = T_{n-1}',$$

then for each stage n,

$$-W_n = C_p T_0 (\alpha_n - 1)$$

and

$$-\Sigma W_n = C_p T_0 \Sigma (\alpha_n - 1).$$

If the α_n in turn are equal and are assigned a value "α_n" where it is required that

$$(\alpha_n)^n = \alpha_n^n = \alpha,$$

then

$$-W_n = C_p T_0 (\alpha^{1/n} - 1)$$

and

$$-\Sigma W_n = C_p T_0 n (\alpha_n - 1) = C_p T_0 n (\alpha^{1/n} - 1).$$

This is as compared to a single-stage compression, where the work term will be

$$-W_{\text{comp}} = C_p T_0 (\alpha - 1).$$

The comparison to a single stage, therefore, involves

$$n(\alpha^{1/n} - 1) \quad \text{versus} \quad (\alpha - 1).$$

For a given α, the term on the left always will be less than the term on the right, for $\alpha > 1$ and $n > 1$. Some comparisons are provided in Table 4.5.

Table 4.5. EFFECT ON COMPRESSION OF THE NUMBER OF STAGES

	$n(\alpha^{1/n} - 1)$		
n	$\alpha = 2$	$\alpha = 3$	$\alpha = 4$
1	1	2	3
2	0.828	1.464	2
3	0.780	1.327	1.762
5	0.744	1.229	1.598
10	0.718	1.161	1.487
100	0.696	1.105	1.396
1000	0.6932	1.0992	1.3873

Infinite Stages. In the limit, for most practical purposes, the exponential can be approximated by a series of the form

$$a^x = 1 + x \ln a + \cdots .$$

Therefore,

$$n(\alpha^{1/n} - 1) \sim n\left(\frac{1}{n}\right)\ln \alpha = \ln \alpha .$$

Thus for large values of n, the ratio $(\ln \alpha)/(\alpha - 1)$ reflects an advantage of multistage compression with interstage cooling. Representative values of the ratio are listed in Table 4.6. The energy savings over single-stage compression progressively increase as α or the compression ratio increases.

At lower compression ratios or values of α, there would be no advantage in interstage cooling because $\ln \alpha \sim (\alpha - 1)$. This lack of advantage applies, in particular, to the Joule cycle with internal heat regeneration,

Table 4.6. COMPARISON BETWEEN INFINITE-STAGE AND SINGLE-STAGE COMPRESSION

α	$\dfrac{\ln \alpha}{(\alpha - 1)}$
1.25	0.893
1.5	0.811
2	0.693
3	0.549
4	0.462
5	0.402
10	0.252

where the efficiency is favored by lower compression ratios or lower values of α.

Heat Exchange. With regard to the heat exchange,

$$Q_0 = 0,$$
$$Q_1 = C_p T_0 (1 - \alpha_1),$$
$$Q_2 = C_p T_0 (1 - \alpha_2),$$
$$\vdots$$
$$Q_{n-1} = C_p T_0 (1 - \alpha_{n-1}),$$
$$Q_n = 0.$$

Observe that in this particular case $Q_0 = 0$ by definition when $T_0 = T_0'$. If the α_n are equal, then ΣQ_n is given by

$$\sum_0^{n-1} Q_i = C_p T_0 (n - 1)(1 - \alpha_n) = C_p T_0 (n - 1)(1 - \alpha^{1/n}),$$

whereas for an adiabatic single-stage compression, $Q_{\text{comp}} = 0$.

It is of interest to note that

$$\Sigma Q_n - \Sigma W_n = C_p T_0 (n - 1)(1 - \alpha^{1/n}) + C_p T_0 n(\alpha^{1/n} - 1)$$
$$= C_p T_0 (\alpha^{1/n} - 1),$$

which is the enthalpy change of the working fluid. For a single-stage adiabatic compression, the enthalpy change is

$$Q_{\text{comp}} - W_{\text{comp}} = C_p T_0 (\alpha - 1).$$

Thus the enthalpy change is much greater for a single-stage compression than for a multistage compression with heat rejection.

Observe that the net enthalpy change in a multistage compression, where the inlet temperature has the same value, is equal to the enthalpy change for the last stage—this is because the enthalpic heat increase is removed or rejected, save for the last stage. In other words, $Q_n = W_n$ except for the last stage.

Isothermal Compression

Note that, as $n \to \infty$,

$$-\Sigma W_n \to C_p T_0 \ln \alpha$$

$$\to C_p T_0 \frac{k-1}{k} \ln\left(\frac{P_n}{P_0}\right),$$

where P_n/P_0 is the overall compression ratio. Therefore,

$$-\Sigma W_n \to C_p T_0 \frac{C_p - C_v}{C_p} \ln\left(\frac{P_n}{P_0}\right) = RT_0 \ln\left(\frac{P_n}{P_0}\right),$$

which is the formula for an isothermal compression at T_0, for example, as used in the derivation for the Carnot cycle.

Similarly, as $n \to \infty$,

$$-\Sigma Q_n \to C_p T_0 \frac{n-1}{n} \ln \alpha \to C_p T_0 \ln \alpha$$

$$\to C_p T_0 \frac{C_p - C_v}{C_p} \ln\left(\frac{P_n}{P_0}\right) = RT_0 \ln\left(\frac{P_n}{P_0}\right).$$

Thus in an isothermal compression, the heat rejected is equal to the work input for compression. The same conclusion can be reached for an isothermal expansion, namely, that the heat input will equal the work done.

Expansion

If, in this case, $\overline{T}_0$ is the constant inlet temperature, then

$$\overline{T}_0' = \overline{T}_1' = \overline{T}_2' = \cdots = \overline{T}_{m-1}'.$$

For each stage m,

$$-\overline{W}_m = C_p \overline{T}_0 \left(\frac{1}{\beta_m} - 1\right)$$

and

$$-\Sigma \overline{W}_m = C_p \overline{T}_0 \Sigma \left(\frac{1}{\beta_m} - 1\right).$$

If the β_m are equal and given a constant value "β_m" whereby

$$(\beta_m)^m = \beta_m^m = \beta,$$

then

$$-\overline{W}_m = C_p \overline{T}_0 \left(\frac{1}{\beta^{1/m}} - 1 \right)$$

and

$$-\Sigma \overline{W}_m = C_p \overline{T}_0 m \left(\frac{1}{\beta_m} - 1 \right) = C_p \overline{T}_0 m \left(\frac{1}{\beta^{1/m}} - 1 \right).$$

For a single-stage expansion, on the other hand, the work term will be

$$-\overline{W}_{\mathrm{exp}} = C_p \overline{T}_0 \left(\frac{1}{\beta} - 1 \right).$$

A comparison will involve the positive terms

$$m \left(1 - \frac{1}{\beta^{1/m}} \right) \quad \text{versus} \quad \left(1 - \frac{1}{\beta} \right).$$

The terms on the left-hand side always will be greater than the term on the right for $\beta > 1$ and $m > 1$. Hence more work of expansion is delivered in the multistage version. Some comparisons are shown in Table 4.7.

Infinite Stages. As m increases without limit, the series approximation is of the form $a^x = 1 + x \ln a + \cdots$ and can be used to obtain

$$m \left(1 - \frac{1}{\beta^{1/m}} \right) \sim m \left(\frac{1}{m} \right) \ln \beta = \ln \beta.$$

Therefore, for larger values of m, the ratio

$$\frac{\ln \beta}{1 - 1/\beta} = \frac{\beta \ln \beta}{\beta - 1}$$

indicates an advantage for multistage expansion with interstage heating as compared to single-stage expansion. Representative values for the ratio are provided in Table 4.8. The work output relative to single-stage expansion increases as β or the expansion ratio increases. For lower values of β or the compression ratio, there would be no advantage.

Table 4.7. EFFECT ON EXPANSION OF THE NUMBER OF STAGES

m	$m(1 - 1/\beta^{1/m})$		
	$\beta = 2$	$\beta = 3$	$\beta = 4$
1	0.5	0.667	0.750
2	0.586	0.845	1.000
3	0.619	0.920	1.110
5	0.647	0.986	1.211
10	0.670	1.040	1.294
100	0.691	1.093	1.377
1000	0.693	1.098	1.385

Heat Exchange. For the heat exchange,

$$Q_0 = 0,$$

$$\overline{Q}_1 = C_p \overline{T}_0 \left(1 - \frac{1}{\beta_2} \right),$$

$$\overline{Q}_2 = C_p \overline{T}_0 \left(1 - \frac{1}{\beta_3} \right),$$

$$\vdots$$

$$\overline{Q}_{m-1} = C_p \overline{T}_0 \left(1 - \frac{1}{\beta_m} \right),$$

$$\overline{Q}_m = 0.$$

Table 4.8. COMPARISON BETWEEN INFINITE-STAGE AND SINGLE-STAGE EXPANSION

β	$\dfrac{\ln \beta}{(1 - 1/\beta)}$
1.25	1.116
1.5	1.2165
2	1.540
3	1.647
4	1.848
5	2.010
10	2.520

Therefore, if the β_m are equal, the sum $\Sigma \overline{Q}_m$ is given by

$$\sum_0^{m-1} \overline{Q}_j = C_p \overline{T}_0 (m-1)\left(1 - \frac{1}{\beta_m}\right) = C_p \overline{T}_0 (m-1)\left(1 - \frac{1}{\beta^{1/m}}\right).$$

For an adiabatic single-stage expansion, $\overline{Q}_{\exp} = 0$.
  Consequently,

$$\Sigma \overline{Q}_m - \Sigma \overline{W}_m = C_p \overline{T}_0 (m-1)\left(1 - \frac{1}{\beta^{1/m}}\right) + C_p \overline{T}_0 m \left(\frac{1}{\beta^{1/m}} - 1\right)$$

$$= -C_p \overline{T}_0 \left(1 - \frac{1}{\beta^{1/m}}\right),$$

which is the enthalpy change of the working fluid, that is, the difference between the heat added to the fluid and the work done by the fluid. For a single-stage adiabatic expansion,

$$\overline{Q}_{\exp} - \overline{W}_{\exp} = -C_p \overline{T}_0 \left(1 - \frac{1}{\beta}\right).$$

Thus the enthalpy change is greater for a single-stage adiabatic expansion than for a multistage expansion with interstage heating. The negative value in both instances denotes the fact that more work is done by the expanding fluid than there is heat added.

  The net enthalpy change for the working fluid in a multistage expansion where the inlet temperatures have the same constant value will be equal to the change for the last stage. In the other stages, the heat added to each stage is equal to the work done at that stage; that is, $\overline{Q}_m = \overline{W}_m$ except for the last stage.

Combined Compression and Expansion

The net work ($-W$, added) is given by the summation

$$-\Sigma W_n - \Sigma \overline{W}_n = C_p T_0 n (\alpha_n - 1) + C_p \overline{T}_0 m \left(\frac{1}{\beta_m} - 1\right)$$

$$= C_p n (\alpha_n - 1)\left(T_0 - \frac{\overline{T}_0}{\beta_m}\right)$$

if $n = m$ and if $\alpha_n = \beta_m$. The net heat added, in turn, becomes

$$\Sigma Q_n + \Sigma \overline{Q}_m = C_p T_0 (n-1)(1-\alpha_n) + C_p \overline{T}_0 (m-1)\left(1 - \frac{1}{\beta_m}\right)$$

$$= -C_p(n-1)(\alpha_n - 1)\left(T_0 - \frac{\overline{T}_0}{\beta_m}\right),$$

where it is also understood that $\alpha_n = \alpha^{1/n}$ and $\beta_m = \beta^{1/m}$. Furthermore,

$$T_n = T_0 \alpha_n \quad \text{and} \quad \overline{T}_0 = \overline{T}_m \beta_m.$$

Under the constraints, therefore,

$$\overline{T}_0 - T_n = \alpha_n(\overline{T}_m - T_0),$$

where expectedly $\overline{T}_0 > T_n$ and $\overline{T}_m > T_0$. Also,

$$\alpha_n = \frac{T_n}{T_0} = \frac{\overline{T}_0}{\overline{T}_m} = \beta_m$$

or

$$\frac{T_n}{\overline{T}_0} = \frac{T_0}{\overline{T}_m}.$$

If $T_n = \overline{T}_0$, then it would be required that $T_0 = \overline{T}_m$ or vice versa.

Cycle Efficiency. When compression and expansion are combined, the cycle efficiency becomes

$$\text{Eff}_{\text{cycle}} = \frac{\text{net work done}}{\text{gross heat added}} = \frac{\Sigma W_n + \Sigma \overline{W}_m}{\Sigma \overline{Q}_m + C_p(\overline{T}_0 - T_n)}$$

$$= \frac{-C_p n(\alpha_n - 1)(T_0 - \overline{T}_0/\beta_m)}{C_p \overline{T}_0 (m-1)(1 - 1/\beta_m) + C_p(\overline{T}_0 - T_n)}.$$

The preceding relationship can be manipulated in various ways. For instance, since $T_n = T_0 \alpha_n$ and if $\alpha_n = \beta_m$, then

$$\text{Eff}_{\text{cycle}} = \frac{n(1 - 1/\alpha_n)(\overline{T}_0 - T_n)}{T_0(n-1)(1 - 1/\alpha_n) + (\overline{T}_0 - T_n)}.$$

Obviously, if $\overline{T}_0 = T_n$, the cycle efficiency becomes zero.

From another standpoint, dividing the numerator and denominator by n,

$$\text{Eff}_{\text{cycle}} = \frac{(1 - 1/\alpha_n)(\bar{T}_0 - T_n)}{((n-1)/n)\bar{T}_0(1 - 1/\alpha_n) + (1/n)(\bar{T}_0 - T_n)}.$$

Alternately,

$$\text{Eff}_{\text{cycle}} = \frac{(\alpha_n - 1)(\bar{T}_m - T_0)}{((n-1)/n)T_m(\alpha_n - 1) + (\alpha_n/n)(\bar{T}_m - T_0)}.$$

The extremes are when $n \to \infty$ and when $n = 1$. Thus when n increases without limit,

$$\text{Eff}_{\text{cycle}} \to \frac{\bar{T}_0 - T_n}{\bar{T}_0} = 1 - \frac{T_n}{\bar{T}_0}$$

$$\to \frac{\bar{T}_m - T_0}{\bar{T}_m} = 1 - \frac{T_0}{\bar{T}_m}.$$

The form has a similarity with the efficiency of the Carnot cycle. For one stage, where $n = 1$,

$$\text{Eff}_{\text{cycle}} = \frac{(1 - 1/\alpha)(\bar{T}_0 - T_n)}{(\bar{T}_0 - T_n)} = 1 - \frac{1}{\alpha}$$

$$= 1 - \frac{T_0}{T_n} = \frac{T_n - T_0}{T_n}$$

$$= 1 - \frac{\bar{T}_m}{\bar{T}_0} = \frac{\bar{T}_0 - \bar{T}_m}{\bar{T}_0}.$$

This is the customary expression for the Joule efficiency. Again, the form has a similarity with the Carnot efficiency.

Observe that the efficiency for $n \to \infty$ is smaller than when $n = 1$:

$$1 - \frac{T_n}{\bar{T}_0} < 1 - \frac{\bar{T}_m}{T_0}$$

or

$$\frac{\overline{T}_0 - T_n}{\overline{T}_0} < \frac{\overline{T}_0 - \overline{T}_m}{\overline{T}_0}$$

if $T_n > \overline{T}_m$. In fact,

$$\text{Eff}_{n \to \infty} = \frac{\overline{T}_0 - T_n}{\overline{T}_0 - \overline{T}_m}\left(1 - \frac{1}{\alpha}\right),$$

where here $\alpha = T_n/T_0 = \overline{T}_0/\overline{T}_m$. Only if $\overline{T}_m > T_n$ could there be any benefit for multistage operation under these conditions. If $T_n = \overline{T}_m$, then the efficiency would be the same as for a single stage.

The foregoing conclusion is apart from fuel efficiency losses, wherein the exit or stack-gas temperature of the fuel combustion products would have to be higher than, say, T_n. Furthermore, heat rejection to the surroundings will limit the temperature level of T_0.

The preceding qualifications severely restrict the value for α, or for the compression–expansion ratio, the same as happens with adiabatic single-stage operations for the Joule cycle.

4.3. NONADIABATIC MULTISTAGE BEHAVIOR WITH CONSTANT INTERSTAGE HEAT TRANSFER

The generalized relationships presented in Section 4.1 convert as follows for constant heat transfer rates between the steps or stages.

Compression

By the notation used,

$$\alpha_n = \frac{T_n}{T'_{n-1}} = \frac{T_n}{T_{n-1} + Q_{n-1}/C_p} = \frac{T_n}{T_{n-1} - a_{n-1}},$$

where Q_{n-1} will have a negative value and a_{n-1} as defined will have a positive value. Therefore,

$$-\frac{Q_{n-1}}{C_p} = a_{n-1} = T_{n-1} - \frac{T_n}{\alpha_n}.$$

Furthermore, if Q_n is a constant, where $n = 0, 1, 2, \ldots, n - 1$, then

$$a = a_0 = a_1 = a_2 = \cdots = a_{n-1} \quad (= a_n)$$

and it follows that

$$\frac{T_0 - T_1}{\alpha_1} = \frac{T_1 - T_2}{\alpha_2} = \frac{T_2 - T_3}{\alpha_3} = \cdots.$$

Assuming that $\alpha_1 = \alpha_2 = \alpha_3 = \cdots = \alpha_n$, then

$$\alpha_n = \frac{T_2 - T_1}{T_1 - T_0} = \frac{\Delta T_2}{\Delta T_1}$$

$$= \frac{T_3 - T_2}{T_2 - T_1} = -\frac{\Delta T_3}{\Delta T_2}$$

$$\vdots$$

$$= \frac{T_n - T_{n-1}}{T_{n-1} - T_{n-2}} = \frac{\Delta T_n}{\Delta T_{n-1}}.$$

Also,

$$\frac{T_n - T_{n-1}}{T_1 - T_0} = \frac{\Delta T_n}{\Delta T_1} = (\alpha_n)^{n-1} = \alpha_n^{n-1}.$$

Alternately,

$$T_1 = T_0 + \Delta T_1,$$

$$T_2 = T_1 + \Delta T_2$$

$$= T_0 + T_1 + \alpha_n \Delta T_1$$

$$= T_0 + (1 + \alpha_n) \Delta T_1,$$

$$T_3 = T_2 + \Delta T_3$$

$$= T_0 + (1 + \alpha_n) \Delta T_1 + \alpha_n^2 \Delta T_1$$

$$= T_0 + (1 + \alpha_n + \alpha_n^2) \Delta T_1,$$

$$\vdots$$

$$T_n = T_0 + (1 + \alpha_n + \alpha_n^2 + \cdots + \alpha_n^{n-1}) \Delta T_1,$$

or

$$\frac{T_n - T_0}{T_1 - T_0} = 1 + \alpha_n + \alpha_n^2 + \cdots + \alpha_n^{n-1}.$$

However,

$$\frac{1}{1 - \alpha_n} = 1 + \alpha_n + \alpha_n^2 + \cdots + \alpha_n^{n-1} + \frac{\alpha_n^n}{1 - \alpha_n}.$$

Therefore,

$$\frac{T_n - T_0}{T_1 - T_0} = \frac{1 - \alpha_n^n}{1 - \alpha_n} = \frac{\alpha_n^n - 1}{\alpha_n - 1} = \frac{\alpha - 1}{\alpha_n - 1}$$

or

$$T_n = T_0 + \frac{\Delta T_1}{\alpha_n - 1}(\alpha_n^n - 1).$$

Observe that

$$T_1 = \alpha_n(T_0 - a_n) = \alpha_n\left(T_0 + \frac{Q_n}{C_p}\right)$$

and

$$\Delta T_1 = T_1 - T_0 = \alpha_n(T_0 - a_n) - T_0 = T_0(\alpha_n - 1) - \alpha_n a_n.$$

The preceding values are therefore affixed by the constant value of $a_n = -Q_n/C_p$, which is chosen.

It may be written that, for the special case where the inlet temperatures are to have the same constant value,

$$T_0 = T_0' = T_1' = T_2' = \cdots = T_{n-1}'.$$

Because the α_n are all equal, then

$$T_1 = T_2 = T_3 = \cdots = T_n.$$

Consequently, for an assigned $a = a_n$ or an assigned $Q = Q_n$,

$$-\frac{Q_n}{C_p} = a_n = T_1\left(1 - \frac{1}{\alpha_n}\right) = T_0(\alpha_n - 1),$$

where Q_n and a_n have constant values. The heat transfer between steps or stages would be designated by these particular constant values.

Expansion

Here,

$$\beta_m = \frac{\overline{T}'_{m-1}}{\overline{T}_m} = \frac{\overline{T}_{m-1} + \overline{Q}_{m-1}/C_p}{\overline{T}_m} = \frac{\overline{T}_{m-1} + b_{m-1}}{\overline{T}_m},$$

where $\overline{Q}_{m-1}$ is positive and b_{m-1} is positive. Therefore,

$$\frac{\overline{Q}_{m-1}}{C_p} = b_{m-1} = \beta_m \overline{T}_m - \overline{T}_{m-1}.$$

If the $\overline{Q}_m$ are constant, where $m = 0, 1, 2, \ldots, m - 1$, then

$$b = b_0 = b_1 = b_2 = \cdots = b_{m-1} \quad (= b_m)$$

and it follows that

$$\beta_1 \overline{T}_1 - \overline{T}_0 = \beta_2 \overline{T}_2 - \overline{T}_1 = \beta_3 \overline{T}_3 - \overline{T}_2 = \cdots.$$

Assuming that $\beta_1 = \beta_2 = \beta_3 = \cdots = \beta_m$, then

$$\frac{1}{\beta_m} = \frac{\overline{T}_2 - \overline{T}_1}{\overline{T}_1 - \overline{T}_0} = \frac{\Delta\overline{T}_2}{\Delta\overline{T}_1}$$

$$= \frac{\overline{T}_3 - \overline{T}_2}{\overline{T}_2 - \overline{T}_1} = \frac{\Delta\overline{T}_3}{\Delta\overline{T}_2}$$

$$\vdots$$

$$= \frac{\overline{T}_m - \overline{T}_{m-1}}{\overline{T}_{m-1} - \overline{T}_{m-2}} = \frac{\Delta\overline{T}_m}{\Delta\overline{T}_{m-1}}$$

and

$$\frac{\overline{T}_m - \overline{T}_0}{\overline{T}_1 - \overline{T}_0} = \frac{\Delta\overline{T}_m}{\Delta\overline{T}_1} = \left(\frac{1}{\beta_m}\right)^{m-1} = \frac{1}{\beta_m^{m-1}}.$$

By the methods derived for compression,

$$\frac{\overline{T}_m - \overline{T}_0}{\overline{T}_1 - \overline{T}_0} = \frac{1 - 1/\beta_m^m}{1 - 1/\beta_m} = \frac{1 - 1/\beta}{1 - 1/\beta_m}$$

or

$$\overline{T}_m = \overline{T}_0 + \frac{\Delta \overline{T}_1}{1 - 1/\beta_m}\left(1 - \frac{1}{\beta_m^m}\right).$$

Note that

$$\overline{T}_1 = \frac{1}{\beta_m}\left(\overline{T}_0 + b\right) = \left(\frac{1}{\beta_m}\right)\left(\overline{T}_0 + \frac{\overline{Q}_m}{C_p}\right)$$

and

$$\Delta \overline{T}_1 = \overline{T}_1 - \overline{T}_0 = \frac{1}{\beta_m}\left(\overline{T}_0 + b_m\right) - \overline{T}_0 = \overline{T}_0\left(\frac{1}{\beta_m} - 1\right) + \frac{b_m}{\beta_m}.$$

The value for $\Delta \overline{T}_1$ will be negative and will be affixed by the constant value chosen for $b = b_m = \overline{Q}_m/C_p$.

For the special case where the inlet temperatures are to have the same constant value,

$$\overline{T}_0 = \overline{T}_0' = \overline{T}_1' = \overline{T}_2' = \cdots = \overline{T}_{m-1}',$$

then, because the β_m are equal,

$$\overline{T}_1 = \overline{T}_2 = \overline{T}_3 = \cdots = \overline{T}_m.$$

Therefore, for an assigned $b = b_m$ or an assigned $\overline{Q} = \overline{Q}_m$,

$$\frac{\overline{Q}_m}{C_p} = b_m = \overline{T}_1 \cdot (\beta_m - 1) = \overline{T}_0\left(1 - \frac{1}{\beta_m}\right).$$

The constant value for $\overline{Q}_m$ represents the heat transfer rate between steps or stages.

Work

The work exchange may be determined by adding the individual terms.

Compression. The terms add as follows for the successive stages:

$$-\Sigma W_n = C_p\left(1 - \frac{1}{\alpha_n}\right)(T_1 + T_2 + \cdots + T_n).$$

Given that

$$T_1 = T_0 + \Delta T_1 \frac{\alpha_n - 1}{\alpha_n - 1},$$

$$T_2 = T_0 + \Delta T_1 \frac{\alpha_n^2 - 1}{\alpha_n - 1},$$

$$T_3 = T_0 + \Delta T_1 \frac{\alpha_n^3 - 1}{\alpha_n - 1},$$

$$\vdots,$$

then

$$T_1 + T_2 + \cdots T_n = nT_0 + \frac{\Delta T_1}{\alpha_n - 1}(\alpha_n - 1 + \alpha_n^2 - 1 + \cdots + \alpha_n^n - 1)$$

$$= nT_0 - n\frac{\Delta T_1}{\alpha_n - 1} + \frac{\Delta T_1}{\alpha_n - 1}(\alpha_n + \alpha_n^2 + \cdots + \alpha_n^n).$$

Furthermore,

$$\frac{\alpha_n}{1 - \alpha_n} = \alpha_n + \alpha_n^2 + \cdots + \frac{\alpha_n^{n+1}}{1 - \alpha_n}.$$

Therefore,

$$T_1 + T_2 + \cdots + T_n = nT_0 - n\frac{\Delta T_1}{\alpha_n - 1} + \frac{\Delta T_1}{\alpha_n - 1}\frac{\alpha_n^{n+1} - \alpha_n}{\alpha_n - 1}$$

$$= nT_0 - \frac{\Delta T_1}{\alpha_n - 1}\left[n - \alpha_n\frac{\alpha_n^n - 1}{\alpha_n - 1}\right]$$

and

$$-\Sigma W_n = C_p\left(1 - \frac{1}{\alpha_n}\right)nT_0 - C_p\,\Delta T_1\left[\frac{n}{\alpha_n} - \frac{\alpha_n^n - 1}{\alpha_n - 1}\right].$$

The temperatures are given by the discrete point behavior

$$T_n = \left(T_0 - \frac{\Delta T_1}{\alpha_n - 1}\right) + \left(\frac{\Delta T_1}{\alpha_n - 1}\right)\alpha_n^n$$
$$= A + B\alpha_n^n,$$

where $n = 1, 2, 3, \ldots$ and A and B are defined in the preceding equation.

Expansion. Here, the work terms add to

$$-\Sigma \overline{W}_m = C_p(1 - \beta_m)\left(\overline{T}_1 + \overline{T}_2 + \cdots + \overline{T}_m\right).$$

Given that

$$\overline{T}_1 = \overline{T}_0 + \Delta\overline{T}_1 \frac{1 - 1/\beta_m}{1 - 1/\beta_m},$$

$$\overline{T}_2 = \overline{T}_0 + \Delta\overline{T}_1 \frac{1 - 1/\beta_m^2}{1 - 1/\beta_m},$$

$$\overline{T}_3 = \overline{T}_0 + \Delta\overline{T}_1 \frac{1 - 1/\beta_m^3}{1 - 1/\beta_m},$$

$$\vdots \,,$$

then

$$\overline{T}_1 + \overline{T}_2 + \cdots + \overline{T}_m = m\overline{T}_0 + \frac{\Delta\overline{T}_1}{1 - 1/\beta_m}$$

$$\times \left(1 - \frac{1}{\beta_m} + 1 - \frac{1}{\beta_m^2} + \cdots + 1 - \frac{1}{\beta_m^m}\right)$$

$$= m\overline{T}_0 + \frac{\Delta\overline{T}_1}{1 - 1/\beta_m}\left[m - \frac{1}{\beta_m}\frac{1 - 1/\beta_m^m}{1 - 1/\beta_m}\right]$$

and

$$-\Sigma\overline{W}_m = C_p(1 - \beta_m)m\overline{T}_0 - C_p \,\Delta\overline{T}_1\left[m\beta_m - \frac{1 - 1\beta_m^m}{1 - 1/\beta_m}\right].$$

The temperatures are given by the discrete point behavior

$$\overline{T}_m = \left(\overline{T}_0 + \frac{\Delta \overline{T}_1}{1 - 1/\beta_m}\right) - \left(\frac{\Delta \overline{T}_1}{1 - 1/\beta_m}\right)\frac{1}{\beta_m^m}$$

$$= \overline{A} + \overline{B}\frac{1}{\beta_m^m},$$

where $\overline{A}$ and $\overline{B}$ are defined by the substitutions and m takes on the values $m = 1, 2, 3, \ldots$.

Heat

The Q_n and $\overline{Q}_m$ have constant values, as do a_n and b_n. The summations are as follows.

Compression. The heat terms sum to

$$\Sigma Q_n = -C_p \Sigma a_n = -C_p n a_n$$

$$= -C_p n \left(T_0 - \frac{T_1}{\alpha_n}\right).$$

Expansion. The heat terms sum to

$$\Sigma \overline{Q}_m = C_p \Sigma b_m = C_p m b_m$$

$$= C_p m \left(\beta_m \overline{T}_1 - \overline{T}_0\right).$$

For the purposes here, let $n = m$ and $\alpha_n = \beta_m$.

Net Work

By adding the work summations previously obtained,

$$-\Sigma W_n - \Sigma \overline{W}_m = C_p\left(1 - \frac{1}{\alpha_n}\right)nT_0 - C_p\,\Delta T_1\left[\frac{n}{\alpha_n} - \frac{\alpha_n^n - 1}{\alpha_n - 1}\right]$$

$$+ C_p(1 - \beta_m)m\overline{T}_0 - C_p\,\Delta\overline{T}_1\left[m\beta_m - \frac{1 - 1/\beta_m^m}{1 - 1/\beta_m}\right],$$

where it may be assumed that $\alpha_n = \beta_m$ and $n = m$. The right-hand side of the preceding equation represents the net work done on the working fluid. By changing the signs, it becomes the net work done by the working fluid.

Efficiency

An efficiency determination will require the use of the net work done by the working fluid. For example, a cycle efficiency can be designated as

$$\text{Eff}_{\text{cycle}} = \frac{\Sigma W_n + \Sigma \overline{W}_n}{\Sigma \overline{Q}_m + C_p\left(\overline{T}_0 - T_n\right)},$$

where expressions for the terms used were attained previously.

The resulting expression for the cycle efficiency as defined in the preceding equation will be overly complicated, especially in the attempt to incorporate heat transfer with the combustion system, in order to determine fuel efficiency.

Fortuitously—as it turns out—a simpler analysis will be provided in the next section, whereby the rates of heat transfer between the steps or stages are made to vary in geometric progression.

4.4. NONADIABATIC MULTISTAGE BEHAVIOR WITH VARIABLE INTERSTAGE HEAT TRANSFER

The previously derived generalized multistage relationships in Section 4.1 transform as follows for variable heat transfer, whereby the successive interstage rates are in geometric progression.

Compression

From the generalized derivations, it can be seen that the following ratios apply for a constant α_n. Thus

$$\frac{W_2}{W_1} = \frac{T_2}{T_1} = R_2 = \frac{T'_1}{T'_0},$$

$$\frac{W_3}{W_2} = \frac{T_3}{T_2} = R_3 = \frac{T'_2}{T'_1},$$

$$\vdots$$

$$\frac{W_n}{W_{n-1}} = \frac{T_n}{T_{n-1}} = R_n = \frac{T'_{n-1}}{T'_{n-2}},$$

and

$$\frac{W_n}{W_1} = \frac{T_n}{T_1} = R_2 R_3 \cdots R_n = \frac{T'_{n-1}}{T'_0}.$$

Note that $1 < R_n < \alpha_n$ is inferred. Furthermore,

$$W_2 = R_2 W_1,$$
$$W_3 = R_3 W_2 = R_3 R_2 W_1,$$
$$\vdots$$
$$W_n = R_n \cdots R_3 R_2 W_1.$$

If the R_n are equal, then

$$W_n = R^{n-1} W_1, \qquad T_n = R^{n-1} T_1, \qquad T'_{n-1} = R^{n-1} T'_0,$$

and

$$\Sigma W_n = (1 + R + R^2 + \cdots + R^n) W_1.$$

Since

$$\frac{1}{1-R} = 1 + R + R^2 + \cdots + R^{n-1} + \frac{R^n}{1-R},$$

then

$$\Sigma W_n = \frac{R^n - 1}{R - 1} W_1$$

or

$$-\Sigma W_n = C_p T_1 \left(1 - \frac{1}{\alpha_n}\right) \frac{R^n - 1}{R - 1}$$

$$= C_p R T_0 \left(1 - \frac{1}{\alpha_n}\right) \frac{R^n - 1}{R - 1} \qquad (T_1 = R T_0)$$

$$= C_p \frac{T_n}{R^{n-1}} \left(1 - \frac{1}{\alpha_n}\right) \frac{R^n - 1}{R - 1}.$$

Also, since

$$Q_0 = C_p T_0 \left(\frac{R_1}{\alpha_n} - 1\right),$$

$$Q_1 = C_p T_1 \left(\frac{R_2}{\alpha_n} - 1\right),$$

$$\vdots,$$

then

$$\frac{Q_1}{Q_0} = R_1 \frac{R_2/\alpha_2 - 1}{R_1/\alpha_1 - 1}$$

$$\vdots$$

If the R_n are equal and the α_n are equal,

$$\frac{Q_1}{Q_0} = R_1 = R,$$

$$\frac{Q_2}{Q_1} = R_2 = R,$$

$$\vdots$$

$$\frac{Q_{n-1}}{Q_{n-2}} = R_{n-1} = R.$$

Furthermore,

$$Q_{n-1} = R^{n-1}Q_0, \qquad T_{n-1} = R^{n-1}T_0,$$

and

$$\Sigma Q_n = \sum_0^{n-1} Q_i = (1 + R + R^2 + \cdots + R^{n-1})Q_0$$

$$= C_p\left(\frac{T_1}{\alpha_n} - T_0\right)\frac{R^n - 1}{R - 1}$$

$$= C_p T_0\left(\frac{R}{\alpha_n} - 1\right)\frac{R^n - 1}{R - 1}.$$

The preceding value will be negative, indicating heat removal or rejection. As a check,

$$\Sigma Q_n - \Sigma W_n = C_p T_0\left[\left(\frac{R}{\alpha_n} - 1\right) + R\left(1 - \frac{1}{\alpha_n}\right)\right]\frac{R^n - 1}{R - 1}$$

$$= C_p T_0[R - 1]\frac{R^n - 1}{R - 1} = C_p T_0(R^n - 1)$$

$$= C_p(R^n T_0 - T_0) = C_p(T_n - T_0).$$

Note that

$$\frac{-\Sigma W_n}{-\Sigma Q_n} = \frac{R(1 - 1/\alpha_n)}{-(R/\alpha_n - 1)} = \frac{R(\alpha_n - 1)}{\alpha_n - R} = \frac{\alpha_n - 1}{\alpha_n/R - 1} > 1;$$

that is, more work is done on the working fluid during compression than there is heat rejected.

Note. As a special case, when $R = 1$, the inlet temperatures become constant, as per Section 4.2.

Expansion

Similar to the derivations presented for compression in the previous section, the following ratios apply to expansion at a constant β_m:

$$\frac{\overline{W}_2}{\overline{W}_1} = \frac{\overline{T}_2}{\overline{T}_1} = S_2 = \frac{\overline{T}_1'}{\overline{T}_0'},$$

$$\frac{\overline{W}_3}{\overline{W}_2} = \frac{\overline{T}_3}{\overline{T}_2} = S_3 = \frac{\overline{T}_2'}{\overline{T}_1'},$$

$$\vdots$$

$$\frac{\overline{W}_m}{\overline{W}_{m-1}} = \frac{\overline{T}_m}{\overline{T}_{m-1}} = S_m = \frac{\overline{T}_{m-1}'}{\overline{T}_{m-2}'},$$

and

$$\frac{\overline{W}_m}{\overline{W}_1} = \frac{\overline{T}_m}{\overline{T}_1} = S_2 S_3 \cdots S_m = \frac{\overline{T}_{m-1}'}{\overline{T}_0'}.$$

Note that it is implied that $S_m < 1$, whereas $\beta_m > 1$. However, $1 < 1/S < \beta_m$.

Furthermore,

$$\overline{W}_2 = S_2 \overline{W}_1,$$

$$\overline{W}_3 = S_3 \overline{W}_2 = S_3 S_2 \overline{W}_1,$$

$$\vdots$$

$$\overline{W}_m = S_m \cdots S_3 S_2 \overline{W}_1.$$

If the S_m are equal, then

$$\overline{W}_m = S^{m-1} \overline{W}_1, \qquad \overline{T}_m = S^{m-1} \overline{T}_1, \qquad \overline{T}_{m-1}' = S^{m-1} \overline{T}_1',$$

and

$$\Sigma \overline{W}_m = (1 + S + S^2 + \cdots + S^m) \overline{W}_1.$$

It follows that

$$\Sigma \overline{W}_n = \frac{S^m - 1}{S - 1} \overline{W}_1 = \frac{1 - S^m}{1 - S} \frac{\overline{W}_m}{S^{n-1}}$$

or

$$-\Sigma \overline{W}_m = C_p \overline{T}_1 (1 - \beta_m) \frac{1 - S^m}{1 - S}$$

$$= C_p S \overline{T}_0 (1 - \beta_m) \frac{1 - S^m}{1 - S} \qquad (\overline{T}_1 = S\overline{T}_0)$$

$$= C_p \frac{\overline{T}_m}{S^{n-1}} (1 - \beta_m) \frac{1 - S^m}{1 - S}.$$

In the same fashion,

$$\overline{Q}_0 = C_p \overline{T}_0 (\beta_1 S_1 - 1),$$

$$\overline{Q}_1 = C_p \overline{T}_1 (\beta_2 S_2 - 1),$$

whereby

$$\frac{\overline{Q}_1}{\overline{Q}_0} = S_1 \frac{\beta_2 S_2 - 1}{\beta_1 S_1 - 1}$$

$$\vdots$$

If the S_m are equal,

$$\frac{\overline{Q}_1}{\overline{Q}_0} = S_1 = S,$$

$$\frac{\overline{Q}_2}{\overline{Q}_1} = S_2 = S,$$

$$\vdots$$

$$\frac{\overline{Q}_{m-1}}{\overline{Q}_{m-2}} = S_{m-1} = S.$$

Moreover,

$$\overline{Q}_{m-1} = S^{m-1} \overline{Q}_0, \qquad \overline{T}_{m-1} = S^{m-1} \overline{T}_0,$$

and

$$\Sigma \overline{Q}_m = \sum_0^{m-1} \overline{Q}_j = (1 + S + S^2 + \cdots + S^{m-1})\overline{Q}_0$$

$$= C_p \left(\beta_m \overline{T}_1 - \overline{T}_0 \right) \frac{1 - S^m}{1 - S}$$

$$= C_p \overline{T}_0 (\beta_m S - 1) \frac{1 - S^m}{1 - S}.$$

The preceding value will be positive.
  As the check,

$$\Sigma \overline{Q}_m - \Sigma \overline{W}_m = C_p \overline{T}_0 (S - 1) \frac{1 - S^m}{1 - S} = C_p \overline{T}_0 (S^m - 1) = C_p \left(\overline{T}_m - \overline{T}_0 \right).$$

Also,

$$\frac{\Sigma \overline{W}_m}{\Sigma \overline{Q}_m} = \frac{S(\beta_m - 1)}{\beta_m S - 1} = \frac{\beta_m - 1}{\beta_m - 1/S} > 1.$$

More work is performed by the working fluid during expansion than there is heat added; that is, $1/S$ is greater than unity.

Overall Temperature Relationships

The overall temperature differences for compression and expansion are related by

$$\overline{T}_0 - T_n = \left(\frac{1}{S} \right)^m \overline{T}_m - R^n T_0.$$

If $1/S = R$ and $m = n$, then the preceding expression simplifies to

$$\overline{T}_0 - T_n = R^n \left(\overline{T}_m - T_0 \right).$$

It also follows that

$$\frac{\overline{T}_0}{T_n} = \frac{(1/S)^m \overline{T}_m}{R^n T_0} = \frac{\overline{T}_m}{T_0} \quad \text{or} \quad \frac{\overline{T}_0}{\overline{T}_m} = \frac{T_n}{T_0}.$$

Net Work Done by the Working Fluid

Adding the work sums yields

$$W_{net} = \Sigma W_n + \Sigma \overline{W}_m$$

$$= -C_p RT_0 \left(1 - \frac{1}{\alpha_n}\right) \frac{R^n - 1}{R - 1} - C_p S \overline{T}_0 (1 - \beta_m) \frac{S^m - 1}{S - 1}.$$

The substitution

$$S \frac{S^m - 1}{S - 1} = S \frac{S^m(1 - 1/S^m)}{S(1 - 1/S)} = S^m \frac{1 - R^n}{1 - R} = S^m \frac{R^n - 1}{R - 1},$$

can be made, where $1/S = R$ and $n = m$. Therefore, whereas $\overline{T}_m = S^m \overline{T}_0$, then

$$W_{net} = C_p \left[\overline{T}_m (\beta_m - 1) - T_0 \frac{R}{\alpha_n} (\alpha_n - 1) \right] \frac{R^n - 1}{R - 1},$$

where $\beta_m = \alpha_n$. Accordingly,

$$W_{net} = \Sigma W_n + \Sigma \overline{W}_m = C_p (\alpha_n - 1) \left[\overline{T}_m - T_0 \frac{R}{\alpha_n} \right] \frac{R^n - 1}{R - 1}.$$

The net work done is positive, which indicates that the work *done* during expansion $(\Sigma \overline{W}_m)$ is greater than the work required for compression $(-\Sigma \overline{W}_n)$.

Note that (R/α_n) is less than unity.

Net Heat Added to the Working Fluid

Adding the sums yields

$$Q_{net} = \Sigma Q_n + \Sigma \overline{Q}_m$$

$$= C_p T_0 \left(\frac{R}{\alpha_n} - 1\right) \frac{R^n - 1}{R - 1} + C_p \overline{T}_0 (\beta_m s - 1) \frac{S^m - 1}{S - 1}.$$

By the previous substitution, where $1/S = R$,

$$Q_{net} = C_p \left[\overline{T}_m \left(\beta_m - \frac{1}{S}\right) - T_0 \frac{1}{\alpha_n} (\alpha_n - R) \right] \frac{R^n - 1}{R - 1},$$

where $\beta_m = \alpha_n$. Therefore,

$$Q_{net} = \Sigma Q_n + \Sigma \overline{Q}_m = C_p(\alpha_n - R)\left[\overline{T}_m - T_0 \frac{1}{\alpha_n}\right]\frac{R^n - 1}{R - 1}.$$

Since $\alpha_n > R$, the net heat added is positive, which indicates that the heat *added* during expansion $(\Sigma \overline{Q}_m)$ is greater than the heat *removed* during compression $(-\Sigma Q_n)$.

Net Heat versus Net Work

Consider the results of the previous determinations, when multiplied out and reduced as follows:

$$Q_{net} - W_{net} = C_p\left[\alpha_n \overline{T}_m - T_0 - R\overline{T}_m + RT_0 \frac{1}{\alpha_n}\right]\frac{R^n - 1}{R - 1}$$

$$- C_p\left[\alpha_n \overline{T}_m - RT_0 - \overline{T}_m + T_0 \frac{R}{\alpha_n}\right]\frac{R^n - 1}{R - 1}$$

$$= C_p\left[T_0(R - 1) - \overline{T}_m(R - 1)\right]\frac{R^n - 1}{R - 1}$$

$$= C_p\left[T_0 - \overline{T}_m\right](R^n - 1) \qquad \overline{T}_m > T_0.$$

For the cycle per se, therefore, more net work would be done than there would be net heat added internally to the working fluid, that is, within the cycle proper. In other words, the enthalpy of the working fluid decreases in going from point T_0 to point $\overline{T}_m$. Additional heat must be supplied from an external source—which will introduce inefficiencies. Note that if $\overline{T}_m = T_0$, then the cycle would ideally be 100% efficient.

As an alternate means of representation,

$$Q_{net} - W_{net} = C_p\left[(T_0 R^n - T_0) - \left(\frac{\overline{T}_m}{S^m} - \overline{T}_m\right)\right]$$

$$= C_p\left[(T_n - T_0) - (\overline{T}_0 - \overline{T}_m)\right],$$

whereby $(\overline{T}_0 - \overline{T}_m)$ is greater than $(T_n - T_0)$.

The foregoing expression also will arrange to

$$Q_{net} - W_{net} = C_p\left[(\overline{T}_m - T_0) - (\overline{T}_0 - T_n)\right],$$

where $\bar{T}_0 - T_n$ is greater than $\bar{T}_m - T_0$. The same result will be obtained by adding the differences

$$\Sigma Q_n - \Sigma W_n \quad \text{and} \quad \Sigma \bar{Q}_m - \Sigma \bar{W}_m.$$

The one result serves as a check on the other.

Relationship between α and R

Consider

$$\alpha_n^n = \alpha, \qquad R^n = \frac{T_n}{T_0}.$$

Therefore,

$$n \ln \alpha_n = \ln \alpha, \qquad n \ln R = \ln\left(\frac{T_n}{T_0}\right),$$

and

$$n = \frac{\ln \alpha}{\ln \alpha_n} = \frac{\ln(T_n/T_0)}{\ln R}.$$

These entities are related through the number of stages.
  More directly,

$$R = \left(\frac{T_n}{T_0}\right)^{1/n}.$$

Thus, given the temperature limits and the number of stages, the required value of R will follow.
  In turn, if $R = 1/S$, then

$$S = \left(\frac{\bar{T}_m}{\bar{T}_0}\right)^{1/m} \quad \text{or} \quad R = \frac{1}{S} = \left(\frac{\bar{T}_0}{\bar{T}_m}\right)^{1/m},$$

where $n = m$.

Maximum Cycle Efficiency

The efficiency for the cycle proper may be determined as follows, first for the case where no rejected heat is recycled, and then for the recycle of the rejected heat.

No Recovery of the Rejected Heat. The efficiency may be calculated from the net work done or else be based on the heat rejected. The same result will be attained. Using the former phrasing,

$$\text{Eff}_{\text{cycle}} = \frac{W_{\text{net}}}{\Sigma \overline{Q}_m + C_p(\overline{T}_0 - T_n)}$$

$$= \frac{C_p(\alpha_n - 1)[\overline{T}_m - T_0(R/\alpha_n)](R^n - 1)/(R - 1)}{C_p\overline{T}_0(\beta_m S - 1)(1 - S^m)/(1 - S) + C_p(\overline{T}_0 - T_n)}.$$

Since

$$S\frac{1 - S^m}{1 - S} = S^m\frac{R^n - 1}{R - 1}, \qquad \overline{T}_m = S^m\overline{T}_0, \qquad n = m, \ \alpha_n = \beta_m, \ R = \frac{1}{S},$$

then, on making the substitutions,

$$\text{Eff}_{\text{cycle}} = \frac{(\alpha_n - 1)[\overline{T}_m - T_0(R/\alpha_n)]}{(\alpha_n - R)\overline{T}_m + (R - 1)/(R^n - 1)(\overline{T}_0 - T_n)}.$$

The maximum efficiency will occur when $\overline{T}_0 - T_n = 0$. This also implies that $\overline{T}_m - T_0 = 0$, because the terms are related by the multiplier $R^n = 1/S^m$.

Therefore, when $\overline{T}_0 = T_n$ and $\overline{T}_m = T_0$,

$$\text{Eff}_{\text{cycle}} = \frac{(\alpha_n - 1)[1 - R/\alpha_n]}{\alpha_n - R}$$

$$= 1 - \frac{1}{\alpha_n}.$$

Other things being equal, this is a relatively low value, because the ordinary Joule cycle will have an efficiency of $1 - 1/\alpha = 1 - 1/(\alpha_n)^n$. On the other hand, due to severe restrictions on the temperature levels of heat rejection, the ordinary Joule cycle can operate only at low values of α. However, this restriction does not apply to the foregoing analysis provided the combustion products are utilized as indicated in Fig. 4.1.

Recovery of Rejected Heat. If the rejected heat is recycled, then the efficiency can be determined as follows. This is an idealization, because no penalty is incurred for raising the rejected heat to a higher temperature level for recycle:

$$\text{Eff}_{\text{cycle}} = \frac{W_{\text{net}}}{Q_{\text{net}} + C_p(\overline{T}_0 - T_n)}$$

$$= \frac{C_p(\alpha_n - 1)[\overline{T}_m - T_0(R/\alpha_n)](R^n - 1)/(R - 1)}{C_p(\alpha_n - R)[\overline{T}_m - T_0(1/\alpha_n)](R^n - 1)/(R - 1) + C_p(\overline{T}_0 - T_n)}$$

$$= \frac{(\alpha_n - 1)[\overline{T}_m - T_0(R/\alpha_n)]}{(\alpha_n - R)[T_m - T_0(1/\alpha_n)] + (R - 1)/(R^n - 1)(\overline{T}_0 - T_n)}.$$

The maximum efficiency occurs when $\overline{T}_0 - T_n = 0$, which infers also that $\overline{T}_m - T_0 = 0$.

The foregoing expression therefore reduces to

$$\text{Eff}_{\text{cycle}} = \frac{(\alpha_n - 1)[1 - R/\alpha_n]}{(\alpha_n - R)[1 - 1/\alpha_n]}$$

$$= \frac{(1 - 1/\alpha_n)[\alpha_n - R]}{(\alpha_n - R)[1 - 1/\alpha_n]}$$

$$= 100\%.$$

The preceding equation does not take into account the energy necessary to recycle the rejected heat. The rejected heat perhaps can be recycled best by using it to preheat the combustion air used to support the overall cycle, as shown in Fig. 4.1.

4.5. NONADIABATIC MULTISTAGE BEHAVIOR SUPPORTED BY COMBUSTION

The scenario using variable interstage heat transfer as presented in Section 4.4 will be adapted to combustive support using the configuration presented in Fig. 4.1. Combustive air, designated as stream W_A, will capture the waste heat rejected from compression and is to be combusted with fuel designated as stream W_F. The combustion products, stream W_B, will be utilized further for interstage heating during expansion.

As a matter of consideration, although the flow diagram shows concurrent flow between the combustion system and the working fluid stream W_{WF}, countercurrent flow preferably should be used in each of the interstage heat exchangers. This will reduce the heat exchanger size requirement, because the mean temperature difference will be greater in countercurrent flow.

Combustion Stoichiometry

If the combustion of the fuel support is represented by

$$C + O_2 \xrightarrow{4N_2} CO_2, \quad \Delta H_{comb} = -168,000 \text{ Btu/lb-mol},$$

then for all practical purposes the molar flow rate of the gaseous combustion system can be regarded as constant. This is not an unreasonable assumption even if natural gas or methane is used as the fuel. Therefore, let

$$W_A = W_B.$$

Furthermore, let $(C_p)_A \sim (C_p)_B$ such that

$$W_A(C_p)_A \sim W_B(C_p)_B.$$

Note that for stoichiometric combustion as in the foregoing representation, $W_A/W_F = 5$. If, however, $W_A/W_F > 5$, only part of the air is used for combustion, whereas if $W_A/W_F < 5$, then there is not sufficient air to support stoichiometric combustion. There is, therefore, the option to introduce additional outside air, say, at the point represented by T_a or T_b. Thus as a minimum requirement, $W_A/W_F = 5$.

Heat Balances

As indicated in Fig. 4.1, it will be assumed that the combustive air, stream A or W_A, picks up the heat rejected during compression. Therefore, per mole of working fluid,

$$W_A(C_p)_A(T_b - T_a) = -\Sigma Q_n$$

$$= -C_p T_0 \left(\frac{R}{\alpha_n} - 1 \right) \frac{R^n - 1}{R - 1},$$

where

$$T_b < T_n, \quad T_a < T_0.$$

Furthermore, during expansion the heat transfer is from the combustion system or combustion products, stream B or W_B, to the working fluid. Thus

$$W_B(C_p)_B(T_c - T_d) = \Sigma \bar{Q}_m$$

$$= C_p \bar{T}_0 (\beta_m S - 1) \frac{S^m - 1}{S - 1}$$

$$= C_p \bar{T}_m (\alpha_n - R) \frac{R^n - 1}{R - 1},$$

where

$$T_c > \bar{T}_0, \qquad T_d > \bar{T}_m,$$

and where

$$R = \frac{1}{S}, \qquad n = m, \qquad \alpha_n = \beta_m, \qquad \bar{T}_m = S^m \bar{T}_0.$$

During combustion it may be assumed that, per unit mole of working fluid,

$$-W_F \Delta H_{comb} - W_A(C_p)_A(T_c - T_b) = C_p(\bar{T}_0 - T_n),$$

where ΔH_{comb} has a negative value, in accordance with the convention. As the limiting case, if $\bar{T}_0 = T_n$, then

$$-W_F \Delta H_{comb} = W_A(C_p)_A(T_c - T_b).$$

Overall, where $W_A(C_p)_A \sim W_B(C_p)_B$, the foregoing expressions also combine to yield

$$-W_F \Delta H_{comb} - W_A(C_p)_A(T_d - T_a)$$

$$= \Sigma Q_n + \Sigma \bar{Q}_m + C_p(\bar{T}_0 - T_n)$$

$$= C_p(\alpha_n - R)\left[\bar{T}_m - T_0 \frac{1}{\alpha_n}\right]\frac{R^n - 1}{R - 1} + C_p(\bar{T}_0 - T_n),$$

where

$$\Sigma Q_n + \Sigma \bar{Q}_m = Q_{net}.$$

As the limiting case, $\bar{T}_0 = T_n$ and $\bar{T}_m = T_0$. Therefore, for the limiting case,

$$-W_F \Delta H_{comb} - W_A(C_p)_A(T_d - T_a) = C_p \bar{T}_m(\alpha_n - R)\left[1 - \frac{1}{\alpha_n}\right]\frac{R^n - 1}{R - 1}.$$

Finally, if $W_A(C_p)_A \sim W_B(C_p)_B$,

$$\frac{T_c - T_d}{T_b - T_a} = -\frac{\overline{T}_m(\alpha_n - R)}{T_0(R/\alpha_n - 1)} = \alpha_n \frac{\overline{T}_m}{T_0}.$$

For the limiting case,

$$\frac{T_c - T_b}{T_b - T_a} = \alpha_n,$$

where $\overline{T}_m = T_0$.

The convention will be adopted that

$$T_0 = T_a + \Delta T_a, \qquad \overline{T}_m = T_d - \Delta T_d,$$

where ΔT_a and ΔT_d signify the respective approaches at the inlet and outlet. The notation also can be used that

$$T_n = T_b + \Delta T_b, \qquad \overline{T}_0 = T_c - \Delta T_c.$$

These approaches may or may not be specified.

Maximum Operating Efficiency

The operating efficiency, on integrating fuel efficiency with cycle efficiency, in general, becomes

$$\text{Eff}_{\text{oper}} = \frac{W_{\text{net}}}{-W_F \Delta H_{\text{comb}}}.$$

The explicit relationship may be obtained by substituting for the entities previously derived. The maximum efficiency, however, will be attained when $\overline{T}_m = T_0$ and $\overline{T}_0 = T_n$. Therefore, with this substitution,

$$\text{Eff}_{\text{oper}} = \frac{C_p(\alpha_n - 1)\overline{T}_m[1 - R/\alpha_n](R^n - 1)/(R - 1)}{C_p(\alpha_n - R)\overline{T}_m[1 - 1/\alpha_n](R^n - 1)/(R - 1) + \Delta_{d-a}}$$

where $\Delta_{d-a} = W_A(C_p)_A(T_d - T_a)$. If $T_d = T_a$,

$$\text{Eff}_{\text{oper}} = \frac{(\alpha_n - 1)[1 - R/\alpha_n]}{(\alpha_n - R)[1 - 1/\alpha_n]} = \frac{(1 - 1/\alpha_n)[\alpha_n - R]}{(\alpha_n - R)[1 - 1/\alpha_n]} = 100\%.$$

This result was obtained also for the ideal case for the cycle efficiency.

Whereas

$$W_A(C_p)_A(T_d - T_c) = -C_p T_0 \left(\frac{R}{\alpha_n} - 1 \right) \frac{R^n - 1}{R - 1} \frac{(T_d - T_a)}{(T_b - T_a)},$$

a major simplification can be effected, whereby

$$\text{Eff}_{\text{oper}} = 1 / \left(1 + \frac{1}{(\alpha_n - 1)} \frac{T_d - T_a}{T_b - T_a} \right).$$

The limiting condition again is satisfied, that the operating efficiency is 100% when $T_d = T_a$.

Note that in the simplification, the following intermediate expression is obtained:

$$\text{Eff}_{\text{oper}} = (1 - R/\alpha_n) / \left[\left(1 - \frac{R}{\alpha_n} \right) + \frac{(1 - R/\alpha_n)}{(\alpha_n - 1)} \frac{T_d - T_a}{T_b - T_a} \right].$$

The factor $1 - R/\alpha_n$ appears in both the numerator and denominator. This brings up the interesting point that any reductions in efficiency caused by mechanical losses or other irreversibilities may cancel out; that is, because the system runs with heat recovery, these effects are recycled from one part of the system to the other—from compression to expansion and vice versa. The net result would be that the overall efficiency tends to be equal to the operating efficiency.

Mechanical Efficiency

The mechanical efficiency may be based on normalizing the compressor and expander efficiencies; that is,

$$\text{Eff}_{\text{mech}} = \frac{\Sigma W_n / E_{\text{comp}} + \Sigma \overline{W}_n (E_{\text{exp}})}{\Sigma W_n + \Sigma \overline{W}_n}.$$

Substituting and simplifying yields

$$\text{Eff}_{\text{mech}} = \frac{\overline{T}_m (E_{\text{exp}}) - (1/E_{\text{comp}}) T_0 (R/\alpha_n)}{\overline{T}_m - T_0 (R/\alpha_n)}.$$

If $\overline{T}_m = T_0$, then

$$\text{Eff}_{\text{mech}} = \frac{E_{\text{exp}} - (1/E_{\text{comp}})(R/\alpha_n)}{1 - R/\alpha_n},$$

where E_{comp} denotes the compressor efficiency and E_{exp} denotes the expander efficiency. For convenience, they may be made equal.

Observe that, because the operating efficiency is independent of the R-value and depends only upon the value of α_n, it is, therefore, possible to lower the R-value and thereby increase the mechanical efficiency; that is, the ratio $T_n/T_0 = \bar{T}_0/\bar{T}_m$ can be reduced, where $R = (T_n/T_0)^{1/n}$. The lower limit, however, is $R = 1$, whereby

$$\text{Eff}_{mech} \rightarrow \frac{E_{exp} - (1/E_{comp})(1/\alpha_n)}{1 - 1/\alpha_n}.$$

This in turn would represent the upper limit for the mechanical efficiency.

Frictional flow losses in the heat exchangers may be included in the mechanical efficiencies also. Of equal interest is the fact that W_{net}, the net work produced per unit of working fluid circulating, will have its largest value for $R = 1$.

Overall Efficiency

The overall efficiency is given by

$$\text{Eff}_{overall} = \text{Eff}_{oper} \cdot \text{Eff}_{mech},$$

where the two terms on the right are determined by the methods previously developed.

Fuel Efficiency. If fuel efficiency is of interest, it can be backed out from the operating efficiency,

$$\text{Eff}_{fuel} = \frac{\text{Eff}_{oper}}{\text{Eff}_{cycle}},$$

where the cycle efficiency is determined as previously shown.

EXAMPLE 4.1:

Helium will be used as the gaseous working fluid, whereby

$$\frac{k-1}{k} = 0.42, \qquad C_p = 4.7 \text{ Btu/lb-mol-°F}.$$

For the combustive air and the combustion products, it will be assumed that

$$(C_p)_A \sim (C_p)_B \sim 7.0.$$

Three each stages will be used for compression and expansion, and each stage will have a compression–expansion ratio of 5. Therefore,

$$n = m = 3$$

and

$$\alpha_n = \beta_m = (5)^{0.42} = \sim 2$$

so that

$$\alpha = (2)^3 = 8$$

and the overall compression–expansion ratio is $5^3 = 125$. Sample calculations will be shown for a 50°F approach at the inlet and at the outlet.

50°F Approach

A 50°F approach will be used at the inlet and outlet, with ambient air at 77°F. Furthermore, the limiting case will be assumed, whereby $\overline{T}_m = T_0$ and $\overline{T}_0 = T_n$. Consequently, let

$$T_a = 77°F \quad (537°R),$$

$$T_0 = 77 + 50 = 127°F \quad (587°R),$$

$$\overline{T}_m = 127°F \quad (587°R),$$

$$T_d = 127 + 50 = 177°F \quad (637°R).$$

Furthermore, assume

$$T_n = 600°F \quad (1060°R),$$

$$\overline{T}_0 = 600°F \quad (1060°R),$$

$$T_b = 600 - 50 = 550°F \quad (1010°R),$$

$$T_b - T_a = 550 - 77 = 473°F \quad (473°R).$$

Thus,

$$R^n = \frac{T_n}{T_0} = \frac{1060}{587} = 1.8058 = \frac{1}{S^m}.$$

For $n = 3$,

$$R = 1.2177, \qquad \frac{R^n - 1}{R - 1} = 3.7014, \qquad \frac{R}{\alpha_n} = 0.6089.$$

Whereas

$$\frac{T_c - T_d}{T_b - T_a} = \alpha_n \quad \text{or} \quad \frac{T_c - 637}{1010 - 537} = 2,$$

then

$$T_c = 1583°R \text{ or } 1123°F.$$

This value will be the (in this case, adiabatic) flame temperature produced. Additionally, it follows that

$$W_A(C_p)_A = \frac{-C_p T_0 (R/\alpha_n - 1)(R^n - 1)/(R - 1)}{T_b - T_a}$$

$$= \frac{-4.7(587)(0.6089 - 1)(3.714)}{437} = 8.444.$$

Therefore,

$$W_A = W_B = 8.444/7.0 = 1.206.$$

These values may be checked from the expression for $W_B(C_p)_B$. Since

$$-W_F \, \Delta H = W_A(C_p)_A(T_c - T_b) = 8.444(1583 - 1010) = 4838.4,$$

then

$$W_F = \frac{4838.4}{168,000} = 0.0288$$

and

$$\frac{W_A}{W_F} = \frac{1.206}{0.0288} = 41.9 > 5.$$

For the maximum operating efficiency, substitution will yield

$$\text{Eff}_{\text{oper}} = \frac{4.7(2 - 1)(587)(1 - 0.6089)(3.7014)}{4.7(2 - 1.2177)(587)(1 - 1/2)(3.7014) + 8.444(100)}$$

$$= \frac{3993.83}{3994.34 + 844.4} = \frac{3993.83}{4838.74} = 82.5\%.$$

For the mechanical efficiency,

$$\text{Eff}_{\text{mech}} = \frac{0.95 - (1/0.95)(0.6089)}{1 - 0.6089} = 79.0\%,$$

where 95% is used for the mechanical efficiency for both compression and expansion. The overall efficiency thereby would be

$$\text{Eff}_{\text{overall}} = 0.825(0.790) = 65.2\%.$$

In this particular instance, the cycle efficiency per se is ideally 100%. The fuel efficiency is, therefore, equal to the operating efficiency.

Note that the operating efficiency more readily can be obtained from the formula

$$\text{Eff}_{\text{oper}} = 1 \Big/ \left(1 + \frac{1}{\alpha_n - 1} \frac{T_d - T_a}{T_b - T_a} \right)$$

$$= 1 \Big/ \left(1 + \frac{1}{2 - 1} \frac{637 - 537}{1010 - 537} \right) = 82.5\%.$$

The operating efficiency is, therefore, independent of R.

The relative magnitude between the energy output from expansion and the energy input for compression is

Expansion	1.0000
Compression	0.6089
Difference	0.3911

The preceding figures indicate why compressor and expander mechanical losses and other irreversibilities can become critical. These losses can cancel out the difference or net energy output.

Comparisons

Some comparisons with other cycles operating at similar temperature levels are in order.

Joule Cycle. If the ordinary Joule cycle could operate at, say, $\alpha = 8$, then the theoretical efficiency would be

$$\text{Eff}_{\text{Joule}} = 1 - \frac{1}{\alpha} = 87.5\%.$$

This figure is strictly hypothetical, because this high a value for α is not feasible using ambient temperatures for heat rejection for both the working fluid and the stack gases. Such a figure would require high exit temperatures for the combustion products or stack gases, which would cause the fuel efficiency to plummet.

Carnot Cycle. For the hypothetical Carnot cycle, adding heat at $T_0 = 1060°R$ and rejecting it at $\overline{T}_m = 587°R$, the efficiency would be

$$\text{Eff}_{\text{Carnot}} = \frac{1060 - 587}{1060} = 44.6\%.$$

Fuel efficiency losses would, of course, reduce this value, because the combustion products or stack gases would have to exit at whatever temperature levels necessary to be compatible with the cycle. The question is academic, however.

Rankine Cycle. Here, the cycle will be assumed to reach temperature levels of 600°F or 1060°R, with the latent heat of condensation rejected, say, at 200°F. Using approximate values for steam, on a unit mass or pound basis, the theoretical efficiency would be roughly

$$\text{Eff}_{\text{Rankine}} \sim \frac{0.5(600 - 200)}{0.5(600 - 200) + 1000} = 16.7\%.$$

The use of enthalpy values from the steam tables will result in a more nearly correct value.

Cost Comparison

For convenience, the cycle under consideration will be labelled the Hoffman cycle. Representative relative values for the expander energy output and the compressor energy input for the cycle are

Expander	1.0000
Compressor	0.6089
Difference	0.3911

whereby the difference or net output is 0.3911 per 1.0000 of expander output.

In terms of 1 megawatt or 1000 kW of net output, the numbers are as follows:

Expander(s)

$$\frac{1}{0.3911}(1000 \text{ kW}) = 2557 \text{ kW},$$

$$(2557)(1.341 \text{ hp/kW}) = 3.403 \text{ hp}.$$

Compressor(s)

$$\frac{0.6089}{0.3911}(1000) = 1557 \text{ kW},$$

$$(1557)(1.341 \text{ hp/kW}) = 2088 \text{ hp}.$$

Representative capital costs based on $100 per horsepower for compression and expansion would be

$$(3403 + 2088)(\$100) = \$549,000.$$

Adding in contingencies, this figure could reach as high as $1 million for the total electrical power generating unit, or approximately the cost of a 1-megawatt gas-fired turbine generator, whereas the cost of a central steam-power plant is circa $1.5 million per megawatt, with everything figured in. Using 10% depreciation, the depreciation costs of a 1-megawatt everything figured in. Using 10% depreciation, the depreciation costs of a 1-megawatt unit may, therefore, be set at $100,000 per year.

Assuming comparable operating costs, exclusive of fuel, for the different methodologies, the comparison, therefore, revolves around fuel costs, which in turn relate to the relative efficiencies. For the purposes here, a figure of 60% will be assigned as compared to a figure of circa 30% at best for gas-fired turbine generators (or the Rankine steam-power cycle).

A comparison per unit megawatt rating for operating over a 330-day year follows.

Fuel Requirement per Year

Theoretical

$$(3412 \text{ Btu/kWh})(1000)(24)(330) = 2.7 \times 10^{10} \text{ Btu}.$$

Hoffman Cycle

$$(2.7 \times 10^{10})\left(\frac{1}{0.60}\right) = 4.5 \times 10^{10}.$$

Gas-Fired Turbine or Rankine Steam Power

$$(2.7 \times 10^{10})\frac{1}{0.30} = 9.0 \times 10^{10}.$$

Total Fuel Costs per Year

Fuel Cost ($/MMBtu)	Theoretical ($)	Hoffman ($)	Gas-fired turbine or Rankine Steam power ($)
1	27,000	45,000	90,000
2	54,000	90,000	180,000
3	81,000	135,000	270,000
4	108,000	180,000	360,000
5	135,000	225,000	450,000

Note that at unit fuel costs as low as $1/MMBtu ($1 per million Btu), the total cost of fuel is of the same order of magnitude as the yearly depreciation cost. At higher unit fuel costs, the savings in total fuel costs using the Hoffman cycle can be more than the yearly plant depreciation costs. Another way of looking at it is that the total fuel costs and savings can be based on a kilowatt-hour of electricity generated; that is, per 330-day year, for a rating of 1-megawatt net output, the total kilowatt-hour of electricity produced is

$$(1000)(24)(330) = 7.92 \times 10^6 \text{ kWh/yr.}$$

The corresponding cost figures per kilowatt-hour are as follows, for the different unit fuel costs and methodologies:

Cost of Fuel per Kilowatt-Hour

Fuel Cost ($/MMBtu)	Theoretical (¢)	Hoffman (¢)	Gas-fired turbine or Rankine Steam Power → (¢)
1	0.34	0.57	1.14
2	0.68	1.14	2.27
3	1.02	1.70	3.41
4	1.36	2.27	4.55
5	1.70	2.84	5.68

As an example, for fuel at $3/MMBtu, there will be a difference of $3.41 - 1.70 = 1.71$¢/kWh between the conventional methodologies and the Hoffman cycle. For electricity ordinarily priced at, say, 5¢/kWh, there would be a cost saving of 34.2%. For electricity nominally at 6¢/kWh, the savings would be 28.5%. Thus savings on the cost of electricity on the order of 30% can be expected. This savings will be magnified as fuel costs increase.

In addition to stretching energy resources, there is a significant cost saving for electrical power generation.

4.6. NONADIABATIC MULTISTAGE BEHAVIOR SUPPORTED BY THERMAL OR GEOTHERMAL ENERGY SOURCES

The methodology for variable interstage heat transfer was presented in Section 4.4 and was adapted to combustive support in Section 4.5. This methodology can also be adapted to exothermic support from chemical and nuclear reactions. The reactants and inerts and the reacting mixture itself, including products, constitute the circulating heat transfer medium, which makes a single pass through the system without recirculation or recycle.

Moreover, there can be sensible and/or latent heat recovery from a liquid or gaseous fluid—in particular, a thermal or geothermal fluid source. That the temperature of a geothermal fluid, for instance, may be relatively low, circa 350°F, is of no theoretical consequence in determining efficiency, which in principle can be much higher than either the Rankin cycle or Carnot cycle. The thermal or geothermal fluid itself becomes the heat transfer medium.

The latter considerations require the recirculation or recycle of a heat transfer medium through the compression and expansion sections, which also may be referred to as the compression and expansion heat transfer sections. A basic flow diagram that is relevant for both a closed-loop and an open-loop flow system is provided in Fig. 4.2. With the closed-loop embodiment, there is no addition or removal of the heat transfer medium. With the open-loop embodiment the fresh heat transfer medium, in part, is fed, introduced, or injected into the recirculating heat transfer medium ahead of the expansion heat transfer section and, in part, is withdrawn or discharged from the recirculating heat transfer medium between the expansion section and the compression section. The remainder of the medium is recirculated through the compression section. Moreover, the discharged medium can be recycled and heated in an external loop and be returned as fresh feed to the system.

A fluid such as a process stream or waste-heat stream, at a temperature sufficiently above ambient conditions, can itself function as both the thermal source and the heat trransfer medium in an open-loop flow system. Part of the cooled fluid is recirculated or recycled through the compressor section.

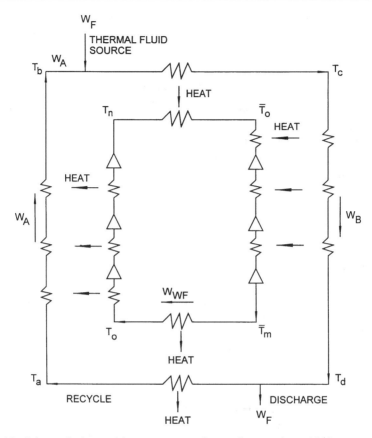

Fig. 4.2. Schematic for multistage compression and expansion with heat recovery using a thermal energy source.

Alternately, another fluid (preferably a liquid) can serve as the heat transfer medium, which is externally heated (countercurrently) by a combustive source or other heat source and fed into the open-loop system as the recirculating heat transfer medium. The heated fluid then acts as a thermal source, and the net effluent discharged after the expansion section can be recycled in an external loop to be reheated by the combustive or other heat source. Although this route is less direct, it will eliminate gas-to-gas heat transfer; instead, there will be gas-to-liquid and liquid-to-gas heat transfer between the stages of compression and the stages of expansion. Brines are, of course, a possibility for the circulating heat transfer medium. If the external loop is closed, other agents such as high-temperature organics (e.g., Dowtherm) or even salt eutectics can be used to sustain elevated temperatures. Some specific examples and applications are as follows.

Geothermal Fluids

As previously indicated, the geothermal fluid itself comprises a recirculating heat transfer medium in an open-loop flow system. After furnishing heat to the expansion section in concurrent flow (as diagrammed in Fig. 4.2), a part of the cooled heat transfer medium is recirculated to remove heat from the compression section; the remaining transfer medium is discharged to the surroundings or discharged back into the geothermal formation.

Streams or Process Streams and Waste-Heat

Any stream or process stream, or waste-heat source, can serve as a thermal source if it is at a sufficient temperature. The stream itself can, in fact, be used as the recirculating heat transfer medium, part or all of which may be recirculated as diagrammed in Fig. 4.2. Alternately, of course, the stream may be used to heat a separate recirculating heat transfer medium by indirect heat transfer, albeit with an expected loss of efficiency.

Waste-Heat from the Rankine Cycle

The waste-heat rejected from the Rankine cycle constitutes a low-grade thermal source. This rejection will occur at the cooler–condenser, where circa two-thirds or more of the heat energy of the fuel used for the Rankine cycle is rejected or lost. This efficiency loss is an unfortunate fact of life for central steam-power plants.

For the most efficient utilization of the waste-heat, the cooler–condenser can be used directly to heat a (liquid) heat transfer medium. Otherwise, the coolant or cooling water can transfer heat to the recirculating heat transfer medium. In fact, the coolant or cooling water itself can become the circulating heat transfer medium.

Stack Gases, Off-Gases, or Combustion Products

The stack gases or off-gases from any combustion process can be used as a thermal energy source. The gases ordinarily have been cooled to circa 300°F, and as such constitute a low-grade waste-heat source. The gases may be used directly as the heat transfer medium, but also may be used to heat a (liquid) heat transfer medium, either ahead of expansion or in an external loop.

Exhaust Gases from the Brayton Cycle

The still-hot exhaust gases from the Brayton cycle can be used as a heat or thermal energy source. The exhaust gases can be introduced ahead of the expansion section and can serve as the recirculating heat transfer medium or they can be used to heat the heat transfer medium ahead of expansion or utilized in an external loop.

Cogeneration or Combined Cycle Power Generation

If two or more power cycles are used in succession—the second working off the first—the overall operation may be referred to as cogeneration or combined cycle power generation. The second cycle with respect to the first may be referred to as a bottoming cycle; the first cycle with respect to the second may be referred to as a topping cycle.

Solar Source

A solar radiant energy source can be utilized to heat the heat transfer medium. The solar radiant energy heats the recirculating heat transfer medium by indirect heat transfer ahead of the expansion section. For continuous nighttime operations and/or on cloudy days, a supplemental heat source can be used, say either thermal or geothermal, or combustive.

Alternately, the cooled heat transfer medium after the expansion section can be recycled and heated in an external loop, with the cooled and heated medium stored in insulated holdup vessels or tanks and in sufficient quantity to maintain continuous operations; that is, to permit a continuation of operations during nighttime and/or on cloudy days.

Nuclear Source

As a special case, nuclear energy can be utilized indirectly as a thermal source; that is, the heat or waste heat generated by nuclear reactions can be used as a heat source. The heat may be added directly to the circulating heat transfer medium via, say, high-pressure–high-temperature water or high-temperature steam produced from the nuclear reactor, or by a high-temperature gas stream (e.g., helium) heated in the nuclear reactor. Furthermore, the high-pressure water, high-temperature steam, or high-temperature gas can serve as the heat transfer medium per se, to be recycled to the reactor in an external loop.

High-Temperature–High-Pressure Water, High-Temperature Steam, or High-Temperature Gases

As outlined in the previous section, high-temperature–high-pressure water can serve as a thermal source. It can be introduced or fed directly to become the recirculating heat transfer medium per se. Alternatively, by indirect heat transfer, it can be used to heat a heat transfer medium ahead of the expansion section or in an external loop.

Similar remarks apply to high-temperature steam, with or without condensation, and to high-temperature gases.

Flow Diagrams, Temperatures, and Material Balances

The basic schematic flow arrangement will be according to Fig. 4.1, with the modifications indicated in Fig. 4.2. The stream designators also represent flow rates and, for the purposes, here will represent molar flow rates.

A cold recycle fluid W_A at temperature T_a is heated to temperature T_b by the heat of compression from successive interstage heat exchangers. The heated fluid is mixed with a hot fluid stream introduced as W_F at a temperature T_F. For the purposes of simplifying the representation, no latent heat effects will be accommodated in the derivations. Furthermore, no chemical reactions or combustion reactions are to occur and no heat of mixing will be assumed. The exothermicity or endothermicity of these additional complications could be included, however, but would unnecessarily complicate the mathematical representation.

The resulting mixture is designated as W_B, where

$$W_A + W_F = W_B.$$

The resulting mixture may or may not be at temperature T_c, depending upon whether a separate heat transfer takes place. For simplification purposes, however, it usually will be assumed as per Fig. 4.1 that

$$T_n = T'_n = \overline{T}_0.$$

The mixture is cooled from T_c to T_d by interstage heat transfer from successive expansions through turbines, to a temperature T_d. The fluid at rate W_B at temperature T_d is, in part, recycled at a rate W_A, with the recycle cooled to T_a; the remainder is discharged at rate W_F at the temperature T_d.

Again for simplicity, it usually will be assumed that

$$T_0 = \overline{T}_m - \overline{T}'_m.$$

The recycled fluid W_A may be cooled by an external air-cooled or water-cooled heat exchanger or cooled in a cooling tower or cooling pond before being returned to the system at the temperature T_a. In the limit, $T_a = T_d$. The necessary derivations are as follows.

Heat Balances

The heat or enthalpy balance for mixing is

$$W_A(C_p)_A(T_b - T_{\text{ref}}) + W_F(C_p)_F(T_F - T_{\text{ref}})$$

$$= W_B(C_p)_B(T_c - T_{\text{ref}}) + C_p(\overline{T}_0 - T_{\text{ref}}) - C_p(T_n - T_{\text{ref}}),$$

where T_{ref} is some arbitrary reference temperature used to calculate values for the corresponding fluid enthalpies. Note that W_A, W_F, and W_B are the molar rates, say, per unit molar rate of the working fluid circulating.

If

$$(C_p)_A \sim (C_p)_F \sim (C_p)_B$$

and if $\overline{T}_0 = T_n$, then

$$W_A T_b + W_F T_F = (W_A + W_F)T_c$$

and

$$\frac{W_F}{W_A} = \frac{T_c - T_b}{T_F - T_c}.$$

This expression relates temperature changes and relative flow rates.

The overall heat balance is

$$W_A(C_p)_A(T_a - T_{\text{ref}}) + W_F(C_p)_F(T_F - T_{\text{ref}})$$

$$= \Sigma Q_n + \Sigma \overline{Q}_m + W_B(C_p)_B(T_d - T_{\text{ref}}).$$

Simplifying,

$$W_A(C_p)_A T_a + W_F(C_p)_F T_F = \Sigma Q_n + \Sigma \overline{Q}_m + W_B(C_p)_B T_d$$

or, since the heat capacities are assumed equal,

$$W_F(C_p)_F(T_F - T_d) = \Sigma Q_n + \Sigma \overline{Q}_m + W_A(C_p)_A(T_d - T_a).$$

Note that the quantity

$$W_F(C_p)_F(T_F - T_d)$$

is the heat given up by the thermal or geothermal source. In turn, the quantity $(C_p)_F(T_F - T_d)$ can be replaced by an enthalpy difference. In this way, latent heat effects—even heats of mixing or heats of reaction—can be incorporated.

Efficiency

The operating efficiency rating may be calculated as

$$\text{Eff}_{\text{oper}} = \frac{W_{\text{net}}}{W_F(T_F - T_d)}$$

or

$$\text{Eff}_{\text{oper}} = \frac{W_{\text{net}}}{\Sigma Q_n + \Sigma \overline{Q}_m + W_A(C_p)_A(T_d - T_a)},$$

where

$$\Sigma Q_n + \Sigma \overline{Q}_m = Q_{\text{net}}$$

and

$$\Sigma W_n + \Sigma \overline{W}_m = W_{\text{net}}.$$

As previously determined in Section 4.4,

$$Q_{\text{net}} - W_{\text{net}} = C_p\left[(T_n - T_0) - (\overline{T}_0 - \overline{T}_m)\right].$$

If $T_n = \overline{T}_0$ and $T_0 = \overline{T}_m$, then

$$Q_{\text{net}} = W_{\text{net}}.$$

Therefore,

$$\text{Eff}_{\text{oper}} = \frac{W_{\text{net}}}{W_{\text{net}} + W_A(C_p)_A(T_d - T_a)}.$$

If $T_d = T_a$, then ideally

$$\text{Eff}_{\text{oper}} = 100\%.$$

This would be the upper theoretical limit.

Determination of W_{net}

In terms of R and S, as also determined in Section 4.4,

$$W_{\text{net}} = -C_p R T_0 \left(1 - \frac{1}{\alpha_n} \right) \frac{R^n - 1}{R - 1} - C_p S T_0 (1 - \beta_m) \frac{S^m - 1}{S - 1}.$$

If $n = m$, $\alpha_n = \beta_m$, and $R = 1/S$, then for the simplification,

$$W_{\text{net}} = -C_p(\alpha_n - 1) \left[\overline{T}_m - T_0 \frac{R}{\alpha_n} \right] \frac{R^n - 1}{R - 1}.$$

The preceding expression may be substituted into the formula for the operating efficiency.

Cold Fluid Rate

It was previously determined (Section 4.5) that

$$W_A (C_p)_A (T_b - T_a) = -\Sigma Q_n$$

$$= -C_p T_0 \left(\frac{R}{\alpha_n} - 1 \right) \frac{R^n - 1}{R - 1},$$

from which W_A can be calculated. Alternately, rewriting the preceding equation and introducing the quantity $T_d - T_a$,

$$W_A (C_p)_A (T_d - T_a) = \frac{T_d - T_a}{T_b - T_a} (-\Sigma Q_n)$$

$$= \frac{T_d - T_a}{T_b - T_a} C_p T_0 \left(1 - \frac{R}{\alpha_n} \right) \frac{R^n - 1}{R - 1}.$$

The preceding expression can, in turn, be used to simplify the efficiency rating.

Simplified Efficiency Rating

Substituting the previously derived expressions into the formula for the operating efficiency rating will achieve the following result:

$$\text{Eff}_{\text{oper}} = \left(C_p(\alpha_n - 1) \left[\overline{T}_m - T_0 \left(\frac{R}{\alpha_n} \right) \right] \frac{R^n - 1}{R - 1} \right) \Bigg/$$

$$\left(C_p(\alpha_n - 1) \left[\overline{T}_m - T_0 \left(\frac{R}{\alpha_n} \right) \right] \frac{R^n - 1}{R - 1} \right.$$

$$+ \frac{T_d - T_a}{T_b - T_a} C_p \left(1 - \frac{R}{\alpha_n} \right) \frac{R^n - 1}{R - 1} \Bigg).$$

If $T_0 = \overline{T}_m$, then

$$\text{Eff}_{\text{oper}} = \frac{\alpha_n - 1}{\alpha_n - 1 + (T_d - T_a)/(T_b - T_a)}.$$

Note that the ratio of the temperature differences will be less than unity because, as expected, $T_d < T_b$. If $T_d = T_a$, the efficiency becomes 100%.

EXAMPLE 4.2:

It will be assumed that the thermal or geothermal source is a water or brine and that helium is the working fluid. As in Example 4.1, it will be stipulated that

$$n = m = 3.$$

The properties of helium are such that

$$\frac{k - 1}{k} = 0.42, \qquad C_p = 4.7 \, \text{Btu/lb-mol-°F}.$$

Furthermore, the compression–expansion ratio is again to be 5, whereby

$$\alpha_n = \beta_m = 5^{0.42} \sim 2$$

so that

$$\alpha = 2^3 = 8$$

and the overall compression–expansion ratio is $5^3 = 125$.

20°F Difference

For illustrative purposes a 20°F difference will be used such that

$$T_d - T_a = 20°F.$$

Additionally it will be assumed that the hot thermal or geothermal source is at a temperature circa $T_F = 350°F$. Accordingly, for a 20°F approach it will be assumed that $T_b = 330°F$. The temperature of the cold recycle fluid will be assumed to be $T_a = 77°F$, whereby for a 20°F difference, $T_d = 97°F$.

Substituting,

$$\text{Eff}_{oper} = \frac{2-1}{2-1 + (97-77)/(330-77)} = 92.7\%.$$

This may be compared to the theoretical results for the Carnot cycle and the Rankine cycle.

Carnot Cycle

The efficiency of a corresponding hypothetical Carnot cycle can be based on the temperature change of the thermal or geothermal fluid. Using this interpretation, with temperatures in degrees absolute,

$$\text{Eff}_{Carnot} = \frac{T_1 - T_2}{T_1} = \frac{T_F - T_d}{T_F} = \frac{(350 + 460) - (97 + 460)}{350 + 460} = 31.2\%.$$

This value is markedly lower than the operating efficiency previously obtained for the nonadiabatic multistage system.

Rankine Cycle

The efficiency of a corresponding Rankine cycle also may be based on the temperature change of the thermal or geothermal fluid or on the temperature change of the working fluid; that is, in the corresponding temperatures used for the Carnot cycle, the Rankine efficiency may be simplified ideally to

$$\text{Eff}_{Rankine} = \frac{T_1 - T_2}{T_1 - T_2 + \lambda/C_p},$$

where λ/C_p is the latent heat of vaporization for the working fluid divided by the heat capacity of the working fluid, in consistent units. This formula may be refined to adjust for actual stream enthalpies or enthalpy differences.

The preceding relationship does not reflect parasitic energy losses, such as for pumping. These losses may be incorporated into the numerator. For the case under consideration, the Rankine efficiency may be rewritten to reflect these losses as

$$\text{Eff}_{\text{Rankine}} = \frac{T_F - T_d - (-\Delta W_{\text{parasitic}})/C_p}{T_F - T_d + \lambda/C_p},$$

where $-\Delta W_{\text{parasitic}}$ denotes the parasitic energy added to the system by the pumping requirements and so forth.

In the case of steam-power generation, where water/steam system is the working fluid, the pumping requirements, say, may be minor as compared to the latent heat change. In geothermal cycles, where, say, isobutane is used as the working fluid, the latent heat change is only about 157 Btu/lb-°F, as compared to up to circa 1000 Btu/lb-°F for steam. Accordingly, for geothermal cycles the parasitic requirements should be taken into account. This may be seen by a comparison of the Carnot and Rankine efficiencies.

Comparison of Carnot and Rankine Efficiencies

The Rankine efficiency can be written as

$$\text{Eff}_{\text{Rankine}} = \frac{T_1 - T_2 - (-\Delta W_{\text{parasitic}})/C_p}{T_1 - (T_2 - \lambda/C_p)}.$$

Accordingly, therefore, if the Rankine efficiency is not to exceed the Carnot efficiency, it is required that

$$\frac{T_1 - T_2}{T_1} > \frac{T_1 - T_2 - (-\Delta W_{\text{parasitic}})/C_p}{T_1 - (T_2 - \lambda/C_p)}.$$

On cross-multiplying, collecting terms, and simplifying, it turns out that

$$\frac{T_1 - T_2}{T_1} > \frac{(-\Delta W_{\text{parasitic}})/C_p}{T_2 - \lambda/C_p}.$$

For isobutane, as the working fluid, in consistent units, $\lambda = 157$ Btu/lb and $C_p = 0.387$ Btu/lb-°F such that $\lambda/C_p = 405.7$.

For the example, T_2 corresponds to $T_d = 97 + 460 = 557$, which is greater than 405.7. It will follow that $(-\Delta W_{\text{parasitic}})$ will have a positive value, whose magnitude is limited by the requirement that the Rankine efficiency does not exceed the Carnot efficiency, which is, manifested in terms of the preceding inequality.

Actual operating data for a conventional Rankine cycle to recover energy from a geothermal fluid may be examined as follows. The information is courtesy of the Barber–Nichols Engineering Company, Arvada, CO.

Isobutane is used as the working fluid. Thermal contact with the geothermal fluid is by indirect heat transfer in a heat exchanger.

The temperature of the hot geothermal brine is 350°F, which is cooled to 110°F and reinjected into the brine-bearing formation at a different location. Condensation takes place in a forced-air air-cooled condenser. The working fluid condenses at 90°F with the ambient air at 70°F. The working fluid condensate is pumped to pressure, then heated, vaporized, and superheated to circa 330°F in the countercurrent heat exchanger. Conversion to electricity is ordinarily estimated at about 95% efficiency.

The turbine output is a nominal 5000 kW, with a generator output of 4656 kW, which here calculates to an electro-mechanical efficiency of about 93%.

The parasitic pumping losses for the isobutane working fluid are 880 kW and the parasitic fan losses for the air-cooled condenser are 562 kW, for a total of 1442 kW. These combined parasitic losses can, therefore, be viewed as about

$$\frac{1442}{5000} = 28.8\%$$

of the heat load. Therefore, for the Rankine efficiency,

$$\text{Eff}_{\text{Rankine}} = \frac{(330 - 110) - [0.288(330 - 110)]/0.387}{(330 - 110) + 157/0.387} = 9.0\%.$$

If the parasitic losses were neglected, the ideal Rankine efficiency would be 35.16%.

Note that

$$\frac{(330 + 460) - (110 + 460)}{330 + 460}$$

$$= 0.278 < \frac{[0.288(330 - 110)]}{(110 + 460) - 157/0.387} = 0.439.$$

In consequence, the actual Rankine efficiency per se would be less than the corresponding Carnot efficiency. The actual Rankine efficiency, in round numbers, is a representative 10%, as determined above.

Also of interest is the comparison of the ideal Rankine efficiency to the corresponding Carnot efficiency. If the latter is to be the greater, then

$$\frac{T_1 - T_2}{T_1} > \frac{T_1 - T_2}{T_1 - (T_2 - \lambda/C_p)}.$$

It follows that

$$T_2 - \frac{\lambda}{C_p} < 0 \quad \text{or} \quad T_2 < \frac{\lambda}{C_p}.$$

This would establish an upper limit for the value of T_2 that can be used, in this case $157/0.387 = 405.7°R$ or $-54.3°F$. Obviously, parasitic effects are of consequences.

Note that for the water/steam system, λ/C_p is on the order of $2000°R$. Hence it is of less consequence to ignore the pumping or parasitic requirements when estimating the Rankine efficiency.

Bottoming versus Topping Cycles

In cogeneration, as it is called, there will be a primary cycle that may be augmented by another cycle ahead of the primary cycle or another cycle aft of the primary cycle. Relevant to the foregoing presentations, by utilizing the waste heat from a primary cycle, such as the Rankine cycle, the afore-described cycle may be referred to as a "bottoming" cycle, as distinguished from a "topping" cycle, which occurs ahead of the primary cycle.

CHAPTER 5

Latent
Heat Recovery

Efficiency savings in power generation and utilization are not the only reasons for advancing the state of technology. While the main emphasis has been on the conversion of the heat of combustion to work to electrical energy, the utilization of thermal and geothermal energy sources has been of concern, as has waste-heat. Thus the previous chapters also dealt with the utilization of waste-heat for power generation, chiefly as low-grade thermal energy in a bottoming cycle. This low-grade thermal energy, for the most part, exists as sensible heat from a single-phase source, which consist of a single component or a mixture. By sensible heat we mean that the heat content is a function of temperature (and pressure) only, and does not involve a change in phase. Or in another way of speaking, we are dealing with a single phase only. Ordinarily, for flow systems, there is a change in temperature with the pressure held constant. There are other instances involving phase changes, however, when it is opportune to recover part of the heat of vaporization or condensation, as the case may be. These phase changes are often referred to as latent heat changes.

In further explanation, the term latent heat or latent heat change denotes a change in enthalpic energy with a change in phase; that is, the system of interest is generally an open or flow system. However, latent changes may also occur in closed systems and are described more appropriately in terms of internal energy; that is, instead of speaking of a latent heat change, we could speak of a latent internal energy change, because "heat" infers enthalpy rather than internal energy and applies to an open or flow system rather than a closed system.

Also to be considered is whether temperature and/or pressure may vary with change of phase. This has to do with the phase rule and the number of components.

To continue, for a single-component system, a change in phase at constant pressure will always occur at constant temperature, whereas for a system of two or more components, a phase change, say, at constant pressure, can also result in a change in temperature; that is, by virtue of the phase rule, the number of degrees of freedom is

$$k - p + 2,$$

where k is the number of components, p is the number of phases, and the 2 represents temperature and pressure. The variables are temperature and pressure, and, for a system of two or more components, the phase compositions. The number of degrees of freedom is equal to the number of *independent* variables.

When the number of degrees of freedom is greater than zero for a fixed number of components and a fixed number of phases, then for each degree of freedom, one of the variables may vary: either phase composition, temperature, or pressure. When the number of degrees of freedom is zero, an invariant point exists and nothing may vary.

At a constant or assigned pressure, for instance, the net number of degrees of freedom is $k - p + 1$, and as long as $k - p + 1 > 0$, there will be at least 1 degree of freedom. For a single-component system at constant pressure the degrees of freedom is $2 - p$. If only one phase exists, the temperature may vary; if two phases coexist, the temperature is fixed.

For a two-component system at constant pressure, there are $3 - p$ degrees of freedom. If one phase exists, both the temperature and a phase composition may vary *independently*. If two phases coexist, the temperature, say, may vary as the independent variable and the phase compositions may vary as dependent variables or vice versa. If three phases coexist, the temperature and phase compositions are fixed. This condition corresponds to a heterogeneous azeotrope in vapor–liquid equilibria or to a eutectic in liquid–solid equilibria.

The foregoing considerations can complicate the analysis of latent heat changes and other energy changes which occur in an open or a closed system. Constraints are incurred involving phase changes and latent heat effects. These complications are above and beyond what is necessary for describing a single-component system undergoing sensible heat changes. The situation is analyzed in reference 1 for heat transfer in condensing pure components and mixtures (e.g., the Nusselt relationships) and also for vaporization.

At constant pressure, for a single-component system, condensation or vaporization will occur at constant temperature. For a mixture or multi-component system, the condensing temperature decreases as condensation proceeds; that is, the less volatile components tend to condense out first. Conversely, during vaporization, the boiling of vaporizing temperature increases as vaporization proceeds because the less volatile components tend to remain in the liquid phase. (Note: for most purposes, condensation and evaporation may be regarded as phenomena occuring at equilibrium.)

Of primary interest here will be the special circumstances under which the heat of compression of a gaseous working fluid can be used as a heat source for drying, evaporation, or vaporization, all of which involve latent heat effects.

Of particular concern is that circumstance in which the system uses what may be called a latent heat pump; that is, the latent heat of condensation of a system is recovered and raised to a higher temperature level in order to effect vaporization or evaporation at the forefront of the same system, or is utilized to vaporize another system (2, 3).

The means for the recovery and utilization of the latent heat from a drying or vaporization system is an ordinary heat pump cycle (or refrigeration cycle). This cycle employs a working fluid, which itself preferably undergoes phase changes; that is, condensed working fluid liquid (WFL) is vaporized when it recovers the heat of condensation at the aft end of the system. The working fluid vapor (WFV) is compressed to a higher temperature, where it is condensed while supplying heat for vaporizing the system liquids at the forefront of the system.

Applications include the drying of solids and the evaporation or vaporization of liquids (from solutions, slurries, or other liquids). Of note are condenser–reboiler coupled distillation columns (4, 5), where the condenser latent heat can be used at the reboiler (if the temperature elevation is not too great).

A special case is that of vapor recompression (6), whereby the vaporized product is compressed to a higher temperature level, followed by heat transfer to the feed in order to cause vaporization or evaporation.

A measure of the effectiveness of latent heat recovery is called the coefficient of performance (COP). The COP increases with the value of the latent heat transferred and will vary inversely with the difference between temperature levels: the larger the temperature difference, the smaller the COP.

These aspects will be examined in the following sections, where the discussion is mainly confined, for obvious reasons, to single-component systems undergoing phase change. Economic comparisons conclude this chapter.

5.1. LATENT HEAT PUMP

The expression latent heat pump will refer to the device or process by which the latent heat recovered from condensation is utilized at a higher temperature for the latent heat of vaporization. The compression of a gaseous working fluid between the associated temperature levels is a principal feature.

To illustrate the principles involved in a latent heat pump for drying, evaporation, or vaporization, a process system for the drying of solids has been selected. In particular, the problem of more efficient removal of free water or moisture from solid fuels such as coal, lignite, or peat, as well as from other wet solids, will be addressed. The system may apply additionally to the removal of such other volatiles as may occur in the temperature range utilized.

The latent heat pump involves the movement and transfer of solids through the system, the transfer of heat to the solids, the evaporation of the volatiles, and the recovery of latent and sensible heats.

A working fluid that optimizes these benefits is utilized in a configuration which also optimizes energy efficiency and operability. While in theory a system could be constructed that reaches 100% recovery, in practical terms there is a trade-off between degree of heat recovery and equipment size and workability. It is the law of diminishing returns which dictates that there be a feasible optimum.

These considerations will be dealt with in detail in the context of this chapter.

Terms

The term latent heat refers to the heat energy required to evaporate water, moisture, or other volatiles. Conversely, the term also serves to denote the heat to be removed during condensation. In either case, there is a phase change involved, and the latent heat denotes the energy associated with the phase change. In the case of evaporation, latent heat may be considered to be a positive term; in condensation it is a negative term.

Sensible heat, on the other hand, refers to the heat required to change the temperature of a gaseous, liquid, or solid phase, most usually at constant pressure (or constant volume). No phase change is involved.

5.2. PROCESS DESCRIPTION

The basic features of the latent heat recovery system are diagrammed in Fig. 5.1. The flow particulars are a wet solids feed whereby the feed is heated and the volatiles are evaporated in heat exchanger A (HT.EX.A), followed by separation and condensation of the volatiles as a separate stream in heat exchanger B (HT.EX.B) along with a cooling of the dried solids. The movement of the solids feed and the dried solids product in and out of the heat exchangers is an important feature of the processing.

In conjunction with this flow there is a working fluid cycle involving the compression (heating) and expansion of a circulating working fluid. The working fluid and cycle conditions are chosen to meet the demands of evaporation and condensation of the moisture and/or other volatiles.

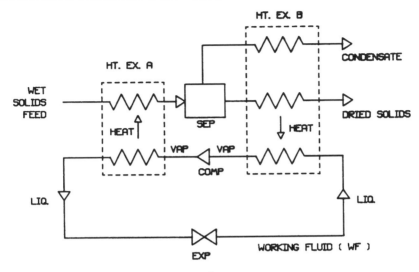

Fig. 5.1. The latent heat recovery system.

A feature of the system is that the working fluid also undergoes phase changes; that is, in heat exchanger A, the heated working fluid vapors from compression are condensed at a higher pressure, whereby the heat of condensation furnishes the sensible heat for heating the solids feed plus the latent heat for vaporizing the moisture or volatiles.

In turn the condensed working fluid liquid leaving heat exchanger A is throttled or expanded to a lower pressure—remaining essentially in the liquid phase. This liquid is vaporized at the lower pressure in heat exchanger B and absorbs the heat necessary to condense the vaporized moisture or volatiles produced from the feed stream (and also to cool the condensate and dried solids).

Fundamentally, the flow diagram in Fig. 5.1 represents the addition of heat to the wet solids feed. This heat is recovered from the condensation of the evaporated moisture product and the cooling of the dried solids product and is recycled to the working fluid.

In further explanation, the heat recycle is accomplished by means of a working fluid (WF) cycle. The working fluid vapor, which may be called a refrigerant, is compressed. Compression raises the working fluid temperature, whereby it transfers heat to the wet solids feed in heat exchanger A. In the process, the working fluid condenses and is further cooled.

The condensed (liquid) working fluid exiting from heat exchanger A is expanded to a lower pressure, a condition at which it will vaporize at a lower temperature, thus retrieving heat from condensing and cooling the products.

The two heat exchangers both involve evaporation and condensation—that is, latent heat changes—but in opposite directions.

Single-Phase Working Fluid

It is also possible to operate the working fluid wholly or partially in the single-phase region as a vapor. However, sensible heat changes would be involved only in the working fluid and would require wide changes in temperature. In other words, the latent heat form is a much more concentrated way to supply and retrieve energy than the sensible heat form.

Furthermore, and equally important, is that at constant pressure the temperature remains constant during the latent heat change; that is, this is true for a single-component working fluid, which will be used here.

Countercurrent Flow

The flow diagram shows the working fluid moving countercurrently to the solids feed. This assures the proper juxtaposition of cooling and heating, and also favors heat transfer; that is, heat exchangers are more effective in countercurrent flow than in concurrent flow (less heat transfer surface area is required for a given set of conditions).

Compression

The compression of a working fluid (in the gaseous state) to raise the temperature also requires a work or energy input, that is, the work of compression. This represents the energy input into the operation to dry the solids. It is the objective, of course, to optimize this work input as related to equipment size and other parameters.

Expansion

The condensed working fluid exists as a liquid phase. It is expanded or throttled to a lower pressure through a valve or choke. This is a so-called isenthalpic expansion, which produces no work. A Pelton wheel or similar device can alternatively be used, and some work can be done or extracted from the expansion. Expansion in the liquid-phase region, however, produces relatively little work.

Another option is to expand the liquid into the vapor region. So-called total flow turbines have been developed which extract work during a two-phase expansion. However, the two-phase expansion will occur at a higher temperature (and pressure), which cuts down the heat transfer effectiveness at heat exchanger B. If the working fluid stayed entirely in the vapor region, then a gaseous expander could be used more effectively to extract energy. Unfortunately, single-phase operation cuts sharply into the effectiveness of heat transfer.

The choice of working fluid and operating pressures assures that the cycle operates in the two-phase region and at temperature levels compatible with the heat transfer requirements.

Block Diagram

In abstract terms, the essence of the process technology is represented by the block diagram that is Fig. 5.2. Aside from the energy required to move the feed and products into, through, and out of the system, the only energy input is the work of compression. The work of compression can be reduced by using a mechanical expander or turbine, for instance; that is,

net work input = work of compression − work of expansion.

The use of an expander turbine may or may not be feasible, however. It will therefore be the principal objective here to relate the net work input to the amount of moisture or volatiles evaporated; in other words, the net energy to evaporate a unit of water, say, British thermal units per pound (Btu/lb).

A measure of the effectiveness of the technology will be how this energy cost (in Btu/lb) compares first to the latent heat of evaporation for water and then to other methods of latent heat recovery such as vapor recompression.

Finally, there is the matter of costs—both fixed and operating—in terms of the moisture reductions.

5.3. THEORY

The theory behind the operation will be derived first in detail. Then such simplifications as are warranted will be made to facilitate calculations and comparisons. The general background was covered in Chapters 1 and 2.

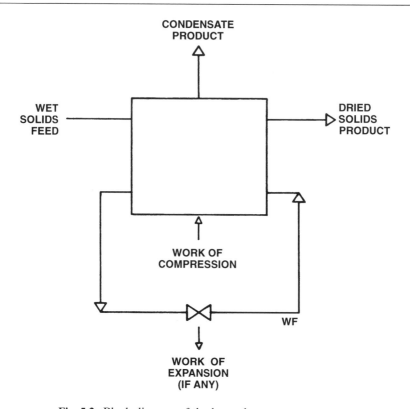

Fig. 5.2. Block diagram of the latent heat recovery system.

The flow diagram is redrawn as represented in Fig. 5.3 to denote different conditions and segments of the system, for the purposes of derivation and calculation. As is usually the case, it is difficult to come up with a uniform set of symbols which are self-evident. The symbolism or nomenclature is as follows:

Feed Side

T_F = temperature of net feed,

T_S = saturation (boiling) temperature of moisture or volatiles,

T_P = temperature at phase separator,

T_C = final temperature of condensate product,

T_D = final temperature of dried product.

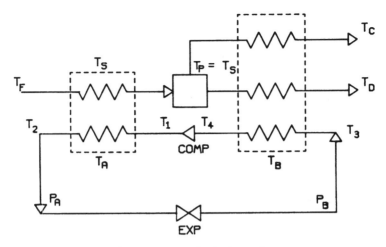

Fig. 5.3. Designated temperatures in the latent heat recovery system.

For simplification it may be assumed that $T_C \sim T_D$. Furthermore, the pressure is assumed uniform on the feed side (at $\sim$ 1-atm pressure).

Working Fluid Side

  T_1 = temperature at compressor outlet and entering heat exchanger A,

  T_2 = temperature leaving heat exchanger A and entering expansion,

  T_3 = temperature leaving expansion and entering heat exchanger B,

  T_4 = temperature leaving heat exchanger B and entering compressor,

  P_A = pressure at compressor outlet and expansion inlet,

  P_B = pressure at compressor inlet and expansion outlet.

The pressure P_A is, therefore, assumed constant over the left leg and P_B is assumed constant over the right leg.

Working Fluid Cycle

The working fluid cycle is diagrammed in pressure–temperature coordinates in Figs. 5.4–5.6. The orientation of the cycle is shown relative to the vapor-pressure curve for the working fluid. The vapor-pressure curve

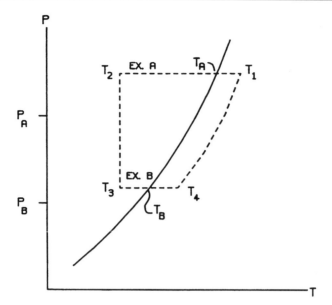

Fig. 5.4. The working fluid is subcooled.

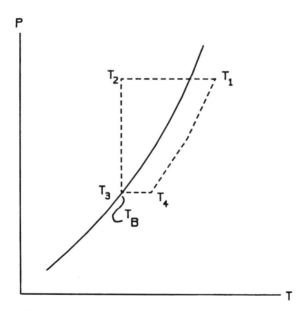

Fig. 5.5. The working fluid reaches the bubble point.

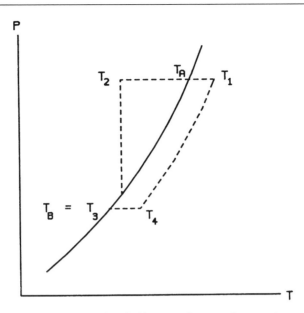

Fig. 5.6. The working fluid enters the two-phase region.

denotes the locus where two phases—vapor and liquid—coexist. Outside of the two-phase locus, a single-phase region(s) exists and is denoted either as liquid or gas. The transition is continuous beyond the critical point.

As the cycle crosses the two-phase locus, condensation or evaporation occurs, depending upon the direction; that is, in heat exchanger A, condensation of the working fluid occurs, whereas in exchanger B, vaporization of the working fluid occurs.

In Fig. 5.4 is shown the cycle whereby sufficient cooling occurs at heat exchanger A such that the liquified working fluid is cooled (supercooled) so as to stay in the liquid region when expanded (T_3). Here $T_2 \sim T_3$.

In Fig. 5.5 the condensed working fluid is expanded to the two-phase region represented by point T_3. Further expansion into the two-phase region will produce cooling, as denoted in Fig. 5.6. The additional cooling will be offset by the subsequent sensible heat requirements for the vapor produced from the expansion (T_3 to T_4).

It is desirable that the locus of compression, denoted by T_4–T_1, stay close to the vapor pressure curve. This will minimize the cooling and heating of the vapor phase. Freons notably have this property.

Conventions

The conventions for heat added to a flow system or feed system F and the work done by the system are diagrammed as follows:

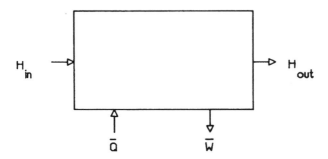

where

H_{in} = unit enthalpy or heat content of stream F entering (Btu/lb or Btu/lb-mol),

H_{out} = unit enthalpy or heat content of stream F leaving (Btu/lb or Btu/lb-mol),

$\overline{Q}$ = heat added to system per unit of F (Btu/lb or Btu/lb-mol),

$\overline{W}$ = work done by system per unit of F (Btu/lb or Btu/lb-mol).

The energy balance is therefore

$$H_{out} - H_{in} = \overline{Q} - \overline{W},$$

where both $\overline{Q}$ and $\overline{W}$ are measured in heat units. (Alternatively, the symbolism $\overline{q}$ and $\overline{w}$ could be used, in work or mechanical units, where

$$\overline{q} = J_0\overline{Q},$$
$$\overline{w} = J_0\overline{W},$$

and J_0 is the mechanical equivalent of heat: 778 ft-lb/Btu.)

The symbols Q and W (or q and w), all without the bar, can be used to denote the total heat added and total work done; that is, for a flow system,

$$F(H_F)_{out} - F(H_F)_{in} = Q - W,$$

where F (or W_F) is the flow rate (in pounds per hour or lb-moles/per hour, say) and H_F corresponds to the unit enthalpy H, but with the designator F added. (The conventions Q' and W' could be used to denote heat rate and work rate, but this would add unnecessarily to the complexity of the symbolism.)

Furthermore, in the directions denoted in the diagram, if the symbol Q (or $\overline{Q}$) is a positive number, this means heat is added to the system; if Q (or $\overline{Q}$) is a negative number, then heat is removed from the system. If the symbol W (or $\overline{W}$) is a positive number, this means work is done by the system; if it is negative, work is done on the system.

Further information on the preceding conventions and the subsequent derivations is contained in Chapters 1 and 2, as previously noted.

Compression

The theoretical relationships for compression—denoted "isentropic" compression—are as follows: for an ideal gas, and in terms of the previous symbols,

$$\frac{T_1}{T_4} = \left(\frac{P_A}{P_B}\right)^{(k-1)/k},$$

where k is the heat capacity ratio $k = C_p/C_v$, C_p is the heat capacity at constant pressure, and C_v is the heat capacity at constant volume. For air, for instance, $k \sim 1.4$ and the exponent is $(k-1)/k \sim 0.28$. Other gases will have different exponents.

The theoretical work becomes

$$-\overline{W}_{\text{theo}} = C_p(T_1 - T_4).$$

This assumes a constant or mean (average) heat capacity. Note that if T_1 is greater than T_4, then $-\overline{W}$ is a positive number or $\overline{W}$ is a negative number that denotes work done on the system.

The actual work done is related to the theoretical by means of an efficiency factor, which also incorporates the nonideality of the system. Thus

$$-\overline{W}_{\text{act}} = \frac{-\overline{W}_{\text{theo}}}{\text{Eff}}.$$

Typical compressor efficiencies may run 80% more or less, depending upon type and conditions.

Expansion

The working fluid ordinarily is expanded or throttled entirely in the liquid region. For all practical purposes, no temperature change occurs and no work is done or incurred. However, other forms of isenthalpic and isentropic expansions can be considered, whereby the working fluid stays in the single-phase gaseous region or whereby the working fluid is condensed and then expanded into the two-phase region. These sorts of derivations are documented elsewhere (e.g., in Chapters 1 and 2).

Material Balance

The material balance for drying the solids is simply

$$W_F = W_C + W_D,$$

where

$$W_F = \text{mass flow rate of wet solids feed,}$$

$$W_C = \text{mass flow rate of condensate product,}$$

$$W_D = \text{mass flow rate of dried solids product.}$$

A mass basis is used herein rather than a molar basis.

Energy Balance

The overall energy balance expressed in terms of unit enthalpies is

$$W_F H_F - W = W_C H_C + W_D H_D,$$

where here $-W$ represents the net work input of compression; that is, W is a negative number. Since $W_F = W_C + W_D$, then

$$-W = W_C(H_C - H_F) + W_D(H_D - H_F).$$

These quantities may further be expressed in terms of temperature as

$$-W_{\text{Theo.}} = W_{\text{WF}}(C_p)_{\text{WFV}}(T_1 - T_4),$$

where W_{WF} is the flow rate of the working fluid. Furthermore,

$$-W_{\text{Act.}} = \frac{-W_{\text{Theo.}}}{\text{Eff}}.$$

For the right-hand side of the energy balance, the terms simplify to

$$W_C(H_C - H_F) = W_C(C_p)_C(T_C - T_F),$$
$$W_D(H_D - H_F) = W_D(Cp)_D(T_D - T_F).$$

Therefore,

$$-W = W_{WF}(C_p)_{WFV}(T_1 - T_4) = W_C(C_p)_C(T_C - T_F) + W_D(C_p)_D(T_D - T_F).$$

If $T_C \sim T_D$, then

$$-W = W_{WF}(C_p)_{WFV}(T_1 - T_4) = \left[W_C(C_p)_C + W_D(C_p)_D\right](T_C - T_F),$$

which establishes the compression requirement.

Heat Transfer Balances

Heat transfer between the process streams and the working fluid involves the following balances, where ΔH is the latent heat for drying and ΔH_{WF} is the latent heat of the working fluid.

Heat Exchanger A (heat load)

$$W_C\left[(C_p)_C(T_S - T_F) + \Delta H\right] + W_D(C_p)_D(T_S - T_F)$$
$$= W_{WF}\left[(C_p)_{WFV}(T_1 - T_A) + \Delta H_{WF} + (C_p)_{WFL}(T_A - T_2)\right].$$

Heat Exchanger B (heat load)

$$W_C\left[\Delta H + (C_p)_C(T_S - T_C)\right] + W_D(C_p)_D(T_S - T_D)$$
$$= W_{WF}\left[(C_p)_{WFL}(T_B - T_3) + \Delta H_{WF} + (C_p)_{WFV}(T_4 - T_B)\right].$$

It is assumed in the foregoing equations that T_S is the saturation temperature at which vaporization of the moisture or volatiles occurs and that no superheating occurs, so that $T_S = T_P$, the temperature of the phase change. As previously noted, it may be assumed also that $T_C = T_D$.

The foregoing information provides a means to relate working fluid rate to other stream rates. Thus

$$W_{WF} = \frac{\left[W_C(C_p)_C(T_S - T_F) + \Delta H\right] + W_D(C_p)_D(T_S - T_F)}{(C_p)_{WFV}(T_1 - T_A) + \Delta H_{WF} + (C_p)_{WFL}(T_A - T_2)};$$

that is, the working fluid flow rate is uniquely determined from the other requirements.

Evaporation

The energy ordinarily required for simple heating followed by evaporation is the sum

$$W_C(C_p)_C(T_S - T_F) + \Delta H + W_D(C_p)_D(T_S - T_F),$$

where T_S is the saturation (or boiling temperature) reached during the evaporation of the moisture or volatiles. In other words, no latent (and sensible heat) recovery is achieved. The foregoing expression may be used as the reference for rating the system effectiveness.

5.4. EFFECTIVENESS RATING

An effectiveness rating may be expressed as the coefficient of performance (COP), defined here as

$$\text{COP}_{\text{theo}} = \frac{W_C(C_p)_C[(T_S - T_F) + \Delta H] + W_D(C_p)_D(T_S - T_F)}{W_{\text{WF}}(C_p)_{\text{WFV}}(T_1 - T_4)}.$$

The numerator is the term for simple evaporation; the denominator is the theoretical (net) work of compression in heat units.
By substituting for W_{WF},

$$\text{COP}_{\text{theo}} = \frac{(C_p)_{\text{WFV}}(T_1 - T_A) + \Delta H_{\text{WF}} + (C_p)_{\text{WFL}}(T_A - T_2)}{(C_p)_{\text{WFV}}(T_1 - T_4)}.$$

If $T_1 = T_A$,

$$\text{COP}_{\text{theo}} = \frac{\Delta H_{\text{WF}} + (C_p)_{\text{WFL}}(T_A - T_2)}{(C_p)_{\text{WFV}}(T_1 - T_4)}.$$

From an inspection of Fig. 5.5, where $T_2 = T_B$

$$T_A - T_2 = T_A - T_B.$$

Therefore,

$$\text{COP}_{\text{theo}} = \frac{\Delta H_{\text{WF}} + (C_p)_{\text{WFL}}(T_A - T_B)}{(C_p)_{\text{WFV}}(T_1 - T_4)}.$$

Furthermore, for Freons,

$$T_1 - T_4 \sim T_A - T_B.$$

Therefore,

$$\text{COP}_{\text{theo}} = \frac{\Delta H_{\text{WF}} + (C_p)_{\text{WFL}}(T_A - T_B)}{(C_p)_{\text{WFV}}(T_A - T_B)}$$

$$= \frac{\Delta H_{\text{WF}}}{(C_p)_{\text{WFV}}(T_A - T_B)} + \frac{(C_p)_{\text{WFL}}}{(C_p)_{\text{WFV}}}.$$

In the instances of interest, the second term on the right-hand side will be small compared to the first term, where T_A and T_B are the saturation temperatures of the working fluid at pressures P_A and P_B, respectively.

Finally, if the theoretical compressor input is to be adjusted for thermoelectric conversions of circa 30%, say, and a compressor efficiency of, say, 80%, the foregoing COP should be multiplied by 0.30(0.80) = 0.24, as follows:

$$\text{COP}_{\text{Act}} = (0.24)\text{COP}_{\text{theo}}$$

$$= (0.24)\frac{\Delta H_{\text{WF}} + (C_p)_{\text{WFL}}(T_A - T_B)}{(C_p)_{\text{WFV}}(T_A - T_B)}$$

$$= (0.24)\frac{\Delta H_{\text{WF}}}{(C_p)_{\text{WFV}}(T_A - T_B)} + \frac{(C_p)_{\text{WFL}}}{(C_p)_{\text{WFV}}}.$$

This will of course markedly reduce the theoretical value.

Boundary Conditions

It is required that

$$T_1 > T_S, \qquad T_3 < T_C \text{ and } T_D,$$
$$T_A > T_S, \qquad T_B < T_S,$$
$$T_2 > T_F, \qquad T_4 < T_S.$$

EXAMPLE 5.1

The following values are specified for a drying operation using Freon-21 as the working fluid:

$$T_A - T_B = 20°F,$$

$$T_S = 212°F,$$

$$\Delta H_{WF} = 87 \text{ Btu/lb for Freon-21},$$

$$(C_p)_{WFV} = 14.6 \text{ Btu/lb-mol for Freon-21},$$

$$(C_p)_{WFL} = 33.98 \text{ Btu/lb-mol for Freon-21},$$

$$(C_p)_{WFL}/(C_p)_{WFV} \sim 2/1,$$

$$MW = 120.94 \text{ for Freon-21}.$$

The value of 20°F for the temperature level difference between the condensation and vaporization of the working fluid is about the reasonable lower limit. Lower values will demand inordinately large exchanger heat transfer surface areas, since the temperature differences to the process vaporization side will be only circa half this, or 10°F.

Based on the preceding figures,

$$COP_{theo} = \frac{87(120.94)}{14.6(20)} + 2$$

$$= 36.03 + 2$$

$$= 38,$$

$$COP_{act} = 0.24(38) = 9.12.$$

These are impressive values for the COP.

Compressor Requirements

The theoretical heat load for the compressor is

$$-W_{theo} = W_C(H_C - H_F) + W_D(H_D - H_F)$$

$$= W_C(C_p)_C(T_C - T_F) + W_D(C_p)_D(T_D - T_F).$$

At 80% efficiency,

$$-W_{act} = \frac{-W_{theo}}{0.80}.$$

The preceding values are in, say, British thermal units per hour (Btu/hr). To convert to compressor horsepower,

$$\text{horsepower}_{\text{act}} = (-W_{\text{act}})(0.0003930).$$

To convert to kilowatts,

$$\text{kilowatts}_{\text{act}} = (-W_{\text{act}})(0.0002930).$$

The cumulative energy requirement in horsepower-hours or kilowatt-hours would be obtained by multiplying by the number of hours of operation.

The theoretical compressor requirement per weight of feed can be expressed as

$$-\overline{W}_{\text{theo}} = \frac{-W_{\text{theo}}}{W_F} = \frac{W_C}{W_F}(C_p)_C(T_C - T_F) + \frac{W_D}{W_F}(C_p)_D(T_D - T_F),$$

where $W_C + W_D = W_F$. Therefore, let

$$W_C + W_D x_2 = W_F x_1 = (W_C + W_D)x_1,$$

where

$$x_1 = \text{weight fraction water in feed,}$$

$$x_2 = \text{weight fraction water left in the dried product.}$$

It follows that

$$W_C[1 - x_1] = W_D[x_1 - x_2]$$

or

$$\frac{W_C}{W_D} = \frac{x_1 - x_2}{1 - x_1}.$$

Furthermore, since $W_F = W_C + W_D$, then

$$\frac{W_F}{W_D} = \frac{W_C}{W_D} + 1$$

or

$$\frac{W_D}{W_F} = \frac{1}{W_C/W_D + 1} = \frac{1}{(x_1 - x_2)/(1 - x_1) + 1} = \frac{1 - x_1}{1 - x_2}.$$

Also

$$\frac{W_F}{W_C} = 1 + \frac{W_D}{W_C}$$

or

$$\frac{W_C}{W_F} = \frac{1}{1 + 1/(W_C/W_D)} = \frac{1}{1 + (1 - x_1)/(x_1 - x_2)} = \frac{x_1 - x_2}{1 - x_2}.$$

Note that

$$\frac{W_D}{W_F} + \frac{W_C}{W_F} = \frac{(1 - x_1) + (x_1 - x_2)}{1 - x_2} = 1.$$

By substituting,

$$-\overline{W}_{\text{theo}} = \frac{x_1 - x_2}{1 - x_2}(C_p)_C(T_C - T_F) + \frac{1 - x_1}{1 - x_2}(C_p)_D(T_D - T_F).$$

The heat capacity of stream D or W_D may be normalized,

$$(C_p)_D = x_2(C_p)_C + (1 - x_2)(C_p)_{\text{mf}},$$

where $(C_p)_{\text{mf}}$ is the heat capacity on a moisture-free basis.

Under typical conditions, assuming a 20°F approach across the system,

$$-\overline{W}_{\text{theo}} = \frac{x_1 - x_2}{1 - x_2}(1.0)(20) + \frac{1 - x_1}{1 - x_2}(0.25)(20)$$

$$= \frac{15x_1 - 20x_2 + 5}{1 - x_2}.$$

The more water left in the product (x_2), the lower the compression requirement.

Working Fluid Rate

The working fluid rate (W_{WF}) is established from

$$-\overline{W}_{\text{theo}}(W_C + W_D) = W_{\text{WF}}(C_p)_{\text{WFV}}(T_1 - T_4),$$

where $T_1 - T_4$ is established from the system characteristics.

Waste-Heat Utilization

If the electrical generating facilities are on site (e.g., a gas-fired turbine cycle), the waste-heat can be used in the drying cycle or used to complement the drying cycle. This waste-heat would amount to perhaps 70% of the fuel burned for electrical power generation. In other words,

$$\text{waste-heat} = \frac{-W_{\text{theo}}}{0.30(0.24)}(0.70)$$

in, say, Btu/hr.

Note. The COP advantage in drying operations can be offset by a large-enough difference between the cost of electrical power for compression and the cost of combustible fuels for direct drying.

5.5. HEAT TRANSFER

Heat transfer in heat exchangers A and B is based on the following equation:

$$\text{heat load} = UA\,\Delta T_{\text{lm}},$$

where

U = overall heat transfer coefficient,

A = heat transfer area,

ΔT_{lm} = log mean temperature difference or equivalent.

For the previous example, the exchanger heat loads are calculated as

$$W_C = \frac{x_1 - x_2}{1 - x_2}W_F = \frac{0.70 - 0.15}{1 - 0.15}\frac{2000(200)}{24} = 10{,}784\ \text{lb/hr},$$

$$W_D = \frac{1 - x_1}{1 - x_2}W_F = \frac{1 - 0.70}{1 - 0.15}\frac{(2000)(200)}{24} = 5882\ \text{lb/hr}.$$

Therefore, the heat loads can be described as follows:

Heat Exchanger A

$$\text{heat load} = W_C\left[(C_p)_C(T_S - T_F) + \Delta H\right] + W_D(C_p)_D(T_S - T_F)$$

$$= 10,784[1.0(212 - 70) + 970]$$

$$\quad + 5882(0.25)(212 - 70)$$

$$= 11.9918 \times 10^6 + 0.2088 \times 10^6$$

$$= 12.2006 \times 10^6 \text{ Btu/hr.}$$

Heat Exchanger B

$$\text{heat load} = W_C\left[\Delta H + (C_p)_C(T_S - T_C)\right] + W_D(C_p)_D(T_S - T_D)$$

$$= 10,784[970 + 1.0(212 - 90)]$$

$$\quad + 5882(0.25)(212 - 90)$$

$$= 11.7761 \times 10^6 + 0.1794 \times 10^6$$

$$= 11.9555 \times 10^6 \text{ Btu/hr.}$$

In theory, the difference between the heat load on heat exchangers A and B should be the theoretical compressor load

$$(12.2006 - 11.9555) \times 10^6 = 245,100 \text{ Btu/hr}$$

Heat Exchanger Sizes

Assuming $T_{lm} = 10°F$ and an overall heat transfer coefficient $U = 100$ Btu/hr-ft^2-°F for each exchanger, the areas can be calculated as follows:

Heat Exchanger A

$$A = \frac{12.2006 \times 10^6}{100(10)} = 12,200 \text{ ft}^2.$$

Heat Exchanger B

$$A = \frac{11.9555 \times 10^6}{100(10)} = 12,000 \text{ ft}^2.$$

Furthermore, heat exchanger B should be in two parts such that the working fluid is divided, with part acting as a separate exchanger for the condensate and part serving for the dried solids. The fraction for partitioning is

$$\frac{0.1794}{11.9555} = 0.015$$

for the solids exchanger. In practice, however, this fraction will be modified by different heat transfer coefficients.

To get an idea of the relative size, a tube which has a diameter of 1 foot will have a surface area of about 3 ft^2/ft. Each 100-ft tube length would therefore have 300 ft of surface. Therefore,

$$\frac{12,000}{300} = 40 \text{ tubes.}$$

That is, 40 tubes, each 1 ft in diameter and 100 ft long, would be required for each exchanger. More and shorter tubes and/or larger diameter tubes would, of course, also serve. An embodiment would require augering or screw-feeding the solids through each tube.

Comment. As will be shown, however, in Section 5.7 on the economic analysis, the compressor electrical requirements are not a major cost item. Hence a larger ΔT and compressor size can be employed to reduce the heat exchanger sizes without significantly affecting plant operating costs.

5.6. VAPOR RECOMPRESSION

The use of vapor recompression is diagrammed in Figure 5.7. The symbols correspond to the previous system, where possible. The cycle is diagrammed in Figure 5.8. Other information about vapor recompression appears, for instance, in reference 6. The analysis follows.

Overall Energy Balance

The overall energy balance in terms of unit enthalpies is

$$W_F H_F - W = W_C H_C + H_D,$$

where $-W$ represents the net theoretical work input of compression.

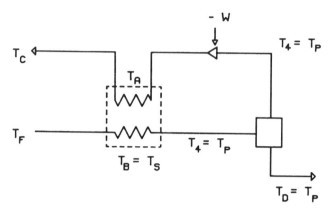

Fig. 5.7. Vapor recompression.

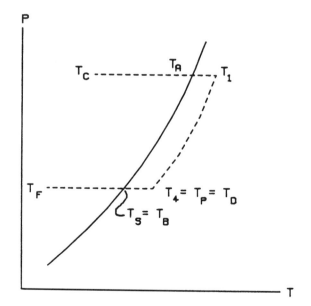

Fig. 5.8. The vapor recompression cycle.

Rearranging, where $W_F = W_C + W_D$, gives

$$-W = W_C(H_C - H_F) + W_D(H_D - H_F)$$
$$= W_C(C_p)_C(T_C - T_F) + W_D(C_p)_D(T_D - T_F).$$

The preceding equation is a theoretical value, from which

$$-W_{act} = \frac{-W_{theo}}{Eff}.$$

The value of $-W_{act}$ is in, say, British thermal units per hour, and may be further converted to horsepower or kilowatts.

Heat Balance at Exchanger

The heat balances are set up in the same manner as before:

$$W_C\left[(C_p)_C(T_S - T_F) + \Delta H + (C_p)_V(T_P - T_S)\right] + W_D(C_p)_D(T_P - T_F)$$
$$= W_C\left[(C_p)_V(T_1 - T_A) + \Delta H + (C_p)_C(T_A - T_C)\right],$$

where V denotes the vapor phase. By rearranging, we obtain

$$W_D(C_p)_D(T_D - T_F) = W_C(C_p)_V[(T_1 - T_A) - (T_P - T_S)]$$
$$+ W_C(C_p)_C[(T_A - T_C) - (T_S - T_F)]$$

or

$$\frac{W_D(C_p)_D}{W_C(C_p)_C} = \frac{(C_p)_V}{(C_p)_C}\frac{(T_1 - T_A) - (T_P - T_S)}{T_D - T_F} + \frac{(T_A - T_C) - (T_S - T_F)}{T_D - T_F}.$$

The ratio of the exiting stream rates is thereby obtained.

Coefficient of Performance

The COP is given by

$$COP_{theo} = \frac{W_C(C_p)_C(T_S - T_C) + \Delta H + W_D(C_p)_D(T_S - T_C)}{W_C(C_p)_V(T_1 - T_P)}.$$

Under the best of circumstances, $T_4 = T_S$ or $T_P = T_S$. However, T_1 has to be greater than T_A since the compression locus diverges away from the vapor–pressure curve, as suggested in Fig. 5.8; that is, with water vapor particularly, the compression curve will have a lesser slope than the vapor–pressure curve in *P-T* space. This is documented in Chapter 6.

Therefore, if $T_P = T_S$, then

$$\text{COP}_{\text{theo}} = \frac{\Delta H}{(C_p)_V (T_1 - T_S)} + \frac{(C_p)_C}{(C_p)_V} \frac{T_S - T_C}{T_1 - T_S} + \frac{W_D (C_p)_D}{W_C (C_p)_C} \frac{T_S - T_C}{T_1 - T_S},$$

where

$$\frac{W_D (C_p)_D}{W_C (C_p)_D} = \frac{(C_p)_V}{(C_p)_C} \frac{T_1 - T_A}{T_P - T_F} + \frac{(T_A - T_C) - (T_S - T_F)}{T_P - T_F}.$$

In final form, where again $T_P = T_S$,

$$\text{COP}_{\text{theo}} = \frac{\Delta H}{(C_p)_V (T_1 - T_S)} + \frac{(C_p)_C}{(C_p)_V} \frac{T_S - T_C}{T_1 - T_S}$$

$$+ \frac{T_S - T_C}{T_1 - T_S} \left\{ \frac{(C_p)_V}{(C_p)_C} \frac{T_1 - T_A}{T_P - T_F} + \frac{(T_A - T_C) - (T_S - T_F)}{T_P - T_F} \right\}.$$

If the second and third terms on the right are negligible, then

$$\text{COP}_{\text{theo}} \sim \frac{\Delta H}{(C_p)_V (T_1 - T_S)}.$$

Finally,

$$\text{COP}_{\text{act}} = (0.24)\text{COP}_{\text{theo}},$$

where the factor 0.24 denotes the effects of thermoelectric conversion and compressor efficiency.

For the case of water, as noted,

$$T_S = T_B = T_4 = T_P = T_D,$$

which will somewhat simplify the expression.

EXAMPLE 5.2

Consider the following conditions for a water–vapor recompression system:

$$T_A = 232°F \; (692°R),$$

$$T_B = T_S = T_4 = T_P = T_D = 212°F \; (672°R),$$

$$T_F = 70°F,$$

$$T_C = 90°F.$$

In degrees absolute, as derived elsewhere (in Section 6.2 of the next chapter), for a diverging compression curve, relative to the vapor–pressure (VP) curve, in log-log coordinates,

$$T_1 = T_B \left(\frac{T_A}{T_B} \right)^{m(k-1)/k},$$

where, for water vapor or steam at these conditions, $k \sim 1.3$ and

$$m = 12.70 \quad \text{(slope of log-log VP curve)},$$

$$\frac{k-1}{k} = \frac{1}{4.17} = 0.24 \quad \text{(inverse of slope of log-log compression curve)}.$$

By substituting, we obtain

$$T_1 = 672 \left(\frac{692}{672} \right)^{12.70/4.17} = 734.78°R \text{ or } 275°F.$$

Furthermore,

$$\Delta H = 970 \text{ Btu/lb},$$

$$(C_p)_V = 0.47 \text{ Btu/lb-°F},$$

$$(C_p)_C = 1.0.$$

Therefore,

$$\text{COP}_{\text{theo}} = \frac{970}{0.47(275 - 212)} + \frac{1.0}{0.47}\frac{212 - 90}{275 - 212}$$

$$+ \frac{212 - 90}{275 - 212}\left\{ \frac{0.47}{1.0}\frac{275 - 232}{212 - 70} + \frac{(232 - 90) - (212 - 70)}{212 - 70}\right\}$$

$$= 32.76 + 4.12 + 1.9365\{0.1423 - 0\}$$

$$= 32.76 + 4.12 + 0.28$$

$$= 37.16.$$

The preceding COP value is approximately the same as for Example 5.1. However, as will be shown subsequently, the base reference requirement for evaporation in vapor recompression is twice that for the latent heat pump.

Comments. The stream W_D exits hot. In principle, this stream could be cooled against incoming feed to further improve the efficiency. However, this would be a solid–solid heat transfer problem, which is not easily resolved.

The stream W_D also could be cooled against the exiting condensate W_C. This, however, would have no effect on the overall efficiency.

Finally, a mitigating argument against drying particulate solids with vapor recompression is the entrainment of solids in the vapor phase. Solids entrainment will adversely affect compressor operation, and the methods necessary to insure solid deentrainment and capture are not viable in a system working under such close temperature and pressure tolerances.

Compressor Requirements (Vapor Recompression)

The theoretical heat load for the compressor is

$$-W_{\text{theo}} = W_C(C_p)_C(T_C - F_F) + W_D(C_p)_D(T_D - T_F),$$

from which it has previously been derived that

$$-\overline{W}_{\text{theo}} = \frac{-W_{\text{theo}}}{W_F} = \frac{x_1 - x_2}{1 - x_2}(C_p)_C(T_C - T_F) + \frac{1 - x_1}{1 - x_2}(C_p)_D(T_D - T_F).$$

Although a 20°F approach may be used for the condensate and feed, the dried product is removed at essentially 212°F. Therefore, assuming $T_F = 70°F$,

$$-\overline{W}_{\text{theo}} = \frac{x_1 - x_2}{1 - x_2}(1.0)(20) + \frac{1 - x_1}{1 - x_2}(0.25)(212 - 70)$$

$$= \frac{-15.5x_1 - 20x_2 + 35.5}{1 - x_2}.$$

The relationship appears somewhat similar to that derived in Section 5.5. By equating the two relationships, it turns out that

$$-15.5x_1 + 35.5 = 15x_1 + 5$$

or

$$30.5 = 30.5x_1.$$

Thus as $x_1 \rightarrow 1.0$, the relationships are identical. Otherwise, for $x_1 < 1.0$, the compressor work requirement for vapor recompression will be higher.

Working Fluid Rate

The working fluid rate (W_{WF}) may be determined from

$$-\overline{W}_{\text{theo}}(W_C + W_D) = W_{\text{WF}}(C_p)_{\text{WFV}}(T_1 - T_4),$$

where $T_1 - T_4 = T_1 - T_S$ is established from the system characteristics.

Comparison

Although the COP can be about the same for vapor recompression as for the previously described technology, this similarity can be misleading because more heat input is required in vapor recompression. This requirement is primarily because the dried solids W_D exit at the drying temperature T_S; that is, for vapor recompression, in terms of the compressor work required,

$$-\overline{W}_{\text{theo}} = \frac{-15.5x_1 - 20x_2 + 35.5}{1 - x_2},$$

whereas for the latent heat pump,

$$-\overline{W}_{theo} = \frac{15x_1 - 20x_2 + 5}{1 - x_2}.$$

The compressor requirement will be considerably lower in the latter case, as has been previously indicated. For instance, consider a peat with 70% moisture reduced to 15% moisture. For vapor recompression,

$$-\overline{W}_{theo} = \frac{-15.5(0.7) - 20(0.15) + 35.5}{1 - 0.15} = 25.5 \text{ Btu/lb},$$

whereas for the latent heat pump,

$$-\overline{W}_{theo} = \frac{15(0.7) - 20(0.15) + 5}{1 - 0.15} = 14.7 \text{ Btu/lb}.$$

In other words, vapor recompression has nearly twice the compression requirement.

In addition, whereas vapor recompression may have twice the ΔT for heat transfer (say, 20°F vs 10°F), the heat load is about twice as great. Thus, effectively the same heat transfer surface area will be involved for vapor recompression as for the latent heat pump system.

The vapor mass flow rate in vapor recompression is W_C. In the example,

$$W_C = 10{,}784 \text{ lb/hr},$$

which is the amount involved in compression.

In the latent heat pump the working fluid mass flow rate is

$$W_{WF} = \frac{-W_{theo}}{(C_p)_{WFV}(T_1 - T_4)}$$

$$= \frac{(14.7 \text{ Btu/lb})(2000)(2000)(1/24)}{(14.6/120.94)(20)}$$

$$= 101{,}747 \text{ lb/hr}.$$

However, vapor recompression operates at 1-atm pressure at 212°F, with a steam (vapor) density of $1/26.80$ lb/ft^3, giving a volumetric rate of

$$10{,}784(26.80) = 289{,}000 \text{ ACF/hr},$$

where ACF denotes actual cubic feet, whereas the heat pump operates at an inlet temperature of 202°F at a pressure of 180 psi, the density is circa 3.07 lb/ft^3 for the Freon-21 working fluid at these conditions, giving a volumetric rate of

$$\frac{101{,}474}{3.07} = 33{,}053 \text{ ACF/hr.}$$

Thus the physical size of the compressor will be much smaller for the latent heat pump system.

5.7. ECONOMICS

The savings generated by the latent heat pump system occur in two forms, as applied, say, to drying coal, lignite, or peat:

1. The extra heat requirement for evaporation of moisture during combustion is avoided.
2. Freight costs are avoided on the moisture or water which is removed.

The analysis of these savings follows.

Evaporative Heat Requirements

Consider a moist or wet fuel with a free moisture content of x_1 that is to be reduced to a fuel of moisture content x_2. Therefore,

$$x_1 = \frac{M_1}{S + M_1},$$

$$x_2 = \frac{M_2}{S + M_2},$$

where

$$M_1 = \text{initial weight of moisture or water,}$$

$$M_2 = \text{final weight of moisture or water,}$$

$$S = \text{weight of fuel solids, a constant.}$$

Solving for M_1 and M_2, we obtain

$$M_1 = \frac{x_1}{1 - x_1} S,$$

$$M_2 = \frac{x_2}{1 - x_2} S.$$

Therefore, the weight of moisture removed is

$$M_1 - M_2 = S\left[\frac{x_1}{1 - x_1} - \frac{x_2}{1 - x_2}\right].$$

Assuming 970 Btu/lb for the latent heat of evaporation of the water removed, the evaporative heat requirement Q is

$$Q = (M_1 - M_2)970 \text{ Btu}$$

$$= S\left[\frac{x_1}{1 - x_1} - \frac{x_2}{1 - x_2}\right]970,$$

where S is measured in pounds.

The unit value of the heat energy required ordinarily is measured in dollars or cents per million Btu or in dollars or cents per Btu. Using \$/MMBtu (dollars per million Btu), the cost of Q Btu's is

$$\text{cost} = Q(\$/\text{MMBtu}) \times 10^{-6}$$

$$= S\left[\frac{x_1}{1 - x_1} - \frac{x_2}{1 - x_2}\right]970(\$/\text{MMBtu}) \times 10^{-6}.$$

This equation gives the cost of removing $M_1 - M_2$ pounds of water using fuel costing \$/MMBtu. The original weight of the wet fuel is $M_1 + S$ pounds. The dried product is $M_2 + S$ pounds. The moisture-free weight is S pounds.

Effect on Evaporation Costs

At this point the reason for using a latent heat pump (LHP) enters; that is, the evaporative heat Q is reduced by the COP factor:

$$Q_{\text{LHP}} = \frac{Q}{\text{COP}}$$

and

$$\text{cost}_{\text{LHP}} = \frac{\text{cost}}{\text{COP}}.$$

Therefore, the evaporative requirement savings on drying becomes

$$\text{savings}_{\text{LHP}} = \text{cost} - \text{cost}_{\text{LHP}}$$

$$= \text{cost}\left[1 - \frac{1}{\text{COP}}\right].$$

This can be a significant quantity.

Comment. Fuels usually are rated according to gross heating value, which is the heating value if the latent heat of vaporization of all the water or moisture initially present and also produced from combustion was recouped by condensation. This is a laboratory measurement, however. In actual practice this moisture or water escapes in the vapor form along with the combustion gases. Therefore, the latent heat of evaporation of the moisture or water in the fuel is not recouped; it represents a heat loss from the heat requirement necessary to vaporize the water during combustion.

Strictly speaking, sensible heat losses also should be included along with the evaporative losses; that is, there is the heat required to heat the moisture to its boiling point ($\sim$ 212°F) plus the heat required to heat the water vapor from the boiling point (212°F) to the stack gas temperature, which may be 300°F or so. Therefore, more strictly speaking, the 970-Btu/lb value used (the latent heat of vaporization) could be adjusted to at least

$$1.0(212 - 100) + 970 + 0.5(300 - 212) = 1126 \text{ Btu/lb},$$

where

$$1.0 = \text{heat capacity of liquid water in Btu/lb-°F},$$

$$0.5 = \text{heat capacity of water vapor (steam)}.$$

For the purposes here the 970-Btu/lb figure will be retained or else it can be rounded off to 1000 Btu/lb.

Freight Adjustment

The rate cost of moving, in this case, solid fuels by rail or truck is in dollars per ton per mile, with the unit rate cost adjusted for mileage, total weight transported, etc. In terms of the water or moisture removed, the equivalent cost of transporting this water, which amounts to a savings, would be

$$\text{savings}_{\text{freight}} = [(M_1 - M_2)/2000](\$/\text{ton-mile})(\text{miles}),$$

where $M_1 - M_2$ is the weight in pounds of the water removed.

Total Savings

The total savings using the latent heat pump (LHP) is, therefore,

$$\text{savings}_{\text{total}} = \text{savings}_{\text{LHP}} + \text{savings}_{\text{freight}},$$

where the quantities on the right-hand side have been defined previously. More precisely, the total savings based on $S + M_1$ pounds of the original wet fuel becomes

$$\text{savings}_{\text{total}} = (M_1 - M_2)\left[970(\$/\text{MMBtu}) \times 10^{-6}\left(1 - \frac{1}{\text{COP}}\right)\right.$$
$$\left. + \frac{(\$/\text{ton-mile})(\text{miles})}{2000}\right],$$

where

$$M_1 - M_2 = S\left[\frac{x_1}{1 - x_1} - \frac{x_2}{1 - x_2}\right].$$

These savings must be compared against the cost of drying by conventional means.

EXAMPLE 5.3

A peat containing 70% moisture is to be dried to 15% moisture. The fuel costs are rated at \$2.50/MMBtu. A COP of 8/1 will be used for the latent heat pump. The dried peat is to be transported 200 miles at a freight rate of \$0.02/ton-mile. The plant capacity is 200 t/day of wet feed (70% moisture). Plant capital costs are an estimated \$800,000 and plant operating costs are an estimated \$80,000 per year.

The cost analysis on the basis of 1 lb of wet feed is as follows:

$M_1 = 0.70$, $x_1 = 0.70$, $(C_p)_C = 1.0$ (water),

$S = 0.30$, $x_2 = 0.15$, $(C_p)_D = 0.25$ (moisture-free peat),

$$M_1 - M_2 = 0.30\left[\frac{0.70}{1 - 0.70} - \frac{0.15}{1 - 0.15}\right] = 0.647 \text{ lb water removed},$$

$$1 - 0.647 = 0.353 \text{ lb of dried product},$$

$$\text{savings}_{\text{LHP}} = 0.647[970(2.50) \times 10^{-6}](1 - \tfrac{1}{8})$$
$$= \$0.001373,$$

$$\text{savings}_{\text{freight}} = 0.647\frac{0.02(200)}{2000}$$
$$= \$0.001294.$$

By adding, we obtain

$$\text{savings}_{\text{total}} = \$0.001373 + \$0.001294 = \$0.002667.$$

Based on 200 t/day, the savings is

$$200(2000)(\$0.002667) = \$1066.80 \text{ per day.}$$

For a 330-day operating year, the savings is

$$330(\$1066.80) = \$352,044 \text{ per year.}$$

The theoretical compression requirement, per pound of net feed and using a 20°F approach, is

$$-\overline{W}_{\text{theo}} = \frac{x_1 - x_2}{1 - x_2}(1.0)(20) + \frac{1 - x_1}{1 - x_2}(0.25)(20)$$

$$= 0.647(20) + 0.353(5)$$

$$= 12.94 + 1.765 - 14.7 \text{ Btu/lb}$$

In terms of kilowatts,

$$14.7(0.0002930) = 0.00431 \text{ kilowatt-hours/lb.}$$

The actual requirement, at 80% efficiency is

$$\frac{0.00431}{0.80} = 0.00539 \text{ kilowatt-hours/lb.}$$

For 330 days of operation per year, the cumulative power is

$$330(24)(0.00539)\frac{200(2000)}{24} = 711,500 \text{ kilowatt-hours.}$$

At 5¢ per kilowatt-hour, the utility costs are

$$711,500(0.05) = \$36,000 \text{ per year.}$$

The compressor size required is

$$\frac{14.7}{0.80}\frac{2000(200)}{24}(0.0003930) = 120 \text{ horsepower.}$$

The electrical rating is

$$120 \frac{(0.000293)}{(0.0003930)} = 89.5 \text{ kilowatts.}$$

The annual plant costs are estimated as follows:

Plant Cost and Depreciation	
$800,000 over 10 years	$ 80,000
Labor and Overhead	
Three men per shift @ $10/hr	80,000
Maintenance	
@ 4%	32,000
Utilities (for compression)	
750,000 kWh @ 5¢	36,000
Peat	
$5/ton (70% moisture)	330,000
	$558,000

The return on investment (ROI) can be based on the savings or based on the expected delivered cost of the dried product.

ROI Based on Savings

$$\text{ROI} = \frac{352,000}{558,000} = 63\%.$$

ROI Based on Dried Product at $15 per ton

$$\text{ROI} = \frac{200(0.353)(330)\$15}{558,000} = 63\%.$$

In other words, if the wet peat could be bought for $5 per ton, dried by the latent heat pump, and the dried product sold at a delivered price of $15 per ton at a terminal 200 miles away, an ROI of 63% could be achieved. That is, the markup is $10 per ton.

The computation can also be prorated based on sale at the mine site (that is, no savings on freight):

ROI Based on Savings

$$\text{ROI} = \frac{352,000(1373/2667)}{558,000} = \frac{181,200}{558,000} = 32.5\%.$$

ROI Based on Dried Product at $7.78 per ton

$$\text{ROI} = \frac{200(0.353)(330)}{558,000} = 32.5\%.$$

In other words, at the mine site the dried product could be sold at $7.78 ton and achieve a 32.5% ROI—a $2.78 markup.

Other comparisons can, of course, be made. It should be noted also that the dried product can be worth as much as $30 per ton delivered.

Note. Since utility costs are a minor part of the total costs, it may not matter how the peat is dried. Note also that at a COP of 8/1, the heat requirements for simple evaporation would be (the ratio of the thermal load to the compressor work load is the COP)

$$14.7(8) = 117.6 \text{ Btu/lb of feed}$$

$$= 1.55 \times 10^{10} \text{ Btu/yr.}$$

At $2.50/MMBtu, this amounts to $38,750, which is about what the utility costs are for compression (at 0.05 per kilowatt-hour). In any event, the energy input costs are relatively small.

As a point of reference, dried peat with a Btu rating of 7000 Btu/lb will have a rating of 14 MMBtu/t. Therefore, at $2.50/MMBtu, the dried product would be worth about $35 per ton.

Additional Note. Capital costs do not reflect the pelletizing of the product. Also, if peat burning facilities were instituted to combust part of the peat for direct drying, pollution controls would become necessary.

Electric Power Generation

Finally, if electric power were generated on the premises (with appropriate pollution controls), there is waste-heat left over which could be used for drying, etc. (There is also the possibility of reducing the equivalent power costs. The electrical power requirement for the foregoing example is 120 hp or 89.5 kW. The relatively small power load, however, probably would not justify setting up a generating unit, even if a small skid-mounted installation. A gas-fired turbine/generator probably would be out of the question for pulverized solid fuels such as peat.) In any event, the total energy input to produce the electrical power requirement (at 80% compressor efficiency and 30% thermoelectric efficiency) would be

$$\frac{14.7(2000)(200)(1.24)}{0.80(0.30)} = 1.02 \times 10^6 \text{ Btu/hr},$$

of which 70% or 0.7×10 Btu/hr is waste heat. Based on the work requirement for compression, which is 14.7 Btu/lb for Example 5.3, the foregoing input can be compared to about

$$14.7(2000)(200)(1/24)(9/1) = 2.2 \times 10^6 \text{ Btu},$$

which would be required for simple evaporation (using a COP of say 9/1, which is the ratio of the thermal load to the work of compression). It could, however, process

$$\frac{0.7}{2.2} \times 200 = 64 \text{ t/day}$$

of wet peat by simple evaporation.

REFERENCES

1. Hoffman, E. J., *Heat Transfer Rate Analysis*, PennWell, Tulsa, OK, 1980.
2. von Cube, H. L. and F. Steiml, *Heat Pump Technology*, I. M. Heinrich, transl., Butterworths, London, 1981, p. 342ff.
3. Heap, R. D., *Heat Pumps*, 2nd ed., E. and F. N. Spon, London, 1983, pp. 180–181.
4. "Proceedings of the First GRI Gas Separations Workshop," sponsored by the Gas Research Institute (Chicago), held at Denver, CO, Oct. 22–23, 1981.
5. Stupin, W. J. and F. J. Lockhart, "Distillation, Thermally Coupled," in *Encyclopedia of Chemical Processing and Design*, Vol. 16, J. J. McKetta and W. Cunningham, eds., Dekker, New York, 1982.
6. Hougen, O. H., K. M. Watson, and R. A. Ragatz, *Chemical Process Principles. II. Thermodynamics*, Wiley, New York, 1959, p. 845ff.

CHAPTER 6

Compression – Expansion
in the Two-Phase Region

As first indicated in Section 1.8 and substantiated in Section 2.2, differential vapor compression along the vapor-pressure curve may result in a smaller work exchange or requirement than for ordinary isentropic compression away from the vapor-pressure curve. At the same time it will be required that there be a heat exchange with or heat rejection to the surroundings or to another medium.

The foregoing will depend upon the characteristics of the working fluid, notably whether the slope of the path of isentropic change is greater or less than the slope of the vapor-pressure curve in, say, pressure-temperature space (or *P-T* space). This not only influences the work requirement, but the heat exchange as well. Similar remarks may be made about expansion.

It is the purpose therefore to examine the behavior more closely when both gaseous and liquid phases are present at saturation, that is, at equilibrium. In further explanation, by saturation we mean a condition that may be represented for most practical purposes as a vapor phase and liquid phase at equilibrium. For a pure substance or compound, the representation is made by means of a vapor-pressure or saturation curve in pressure-temperature space (or *P-T* space), terminating at the critical point. By *P-T* space will be meant that pressure is the ordinate and temperature is the abscissa.

For a mixture, the curve is instead an envelope in *P-T* space, one locus denoting the saturated vapor phase, the other envelope the saturated liquid phase. The loci meet at the critical point for the mixture. The vapor phase may be regarded as at its "dew point," the liquid phase at its "bubble point." The foregoing is all part of the subject of heterogeneous phase equilibria for pure components and mixtures.

Strictly speaking, moreover, the terms vapor and liquid pertain to a system existing at a condition of phase equilibrium, where the less dense phase is called the vapor and the more dense phase is called the liquid—the latter as distinguished from a solid phase. Moreover, the term gas or gaseous phase may be regarded as a superheated vapor. That is, it exists at conditions away from the vapor-pressure curve or saturation curve, or away from the two-phase envelope for a mixture, the representations for vapor-liquid equilibrium behavior at varying pressure and temperature.

281

Furthermore, what we regularly think of as a liquid is, more properly, actually a supercooled liquid, that is, the liquid phase at a condition away from the vapor-pressure curve. The terms are, however, used interchangeably in each case. Interestingly, what we think of as a vapor phase may become a liquid phase without ever undergoing a phase change, or vice versa, by the process of circumventing or circumnavigating the critical point.

The term "wet compression" is sometimes used if a condensate phase is produced or exists, whereas "two-phase" expansion is probably more appropriate for the opposite circumstance. Furthermore there is the likelihood that a phase change is occurring between the gas or vapor and the liquid phase present. That is, either condensation or vaporization may occur, depending upon the vapor-pressure and isentropic characteristics of the working fluid.

Due to equipment problems, the continued presence of a liquid phase during compression is usually considered undesirable, necessitating its removal as compression proceeds. Thus wet compression can be regarded as a process of compressing a vapor at saturation, whereby the vapor rate will change due to partial condensation. In other circumstances, depending upon the nature of the working fluid, injected liquid may be vaporized by the heat of compression. It depends upon the isentropic behavior relative to the vapor-pressure curve.

Similarly, a phase change will expectedly occur during two-phase expansion along the vapor-pressure curve, and may be considered from the standpoint of either an isenthalpic change or an isentropic change. (An isenthalpic expansion is sometimes known as throttling, and is related to Joule-Thomson effects for a single phase—but will also entail overriding latent heat effects if a phase change occurs.) For an isenthalpic change, as the pressure is reduced, further vaporization will occur. As a limiting condition, a liquid may be expanded into the two-phase region, and then into the superheated vapor region. For isentropic-type expansions, which are accompanied by a work exchange, the situation is more involved.

Depending upon the isentropic behavior of the working fluid relative to the vapor-pressure curve, an isentropic single-phase expansion may reach the vapor-pressure curve and enter the two-phase region, producing condensate. It may be observed, therefore, that the presence of a liquid phase in ordinary turbine operation is bad news, unless the equipment is specifically designed for two-phase flow. If this becomes a problem, the expansion can be staged with the liquid condensate removed between stages, and the vapor re-superheated for the next stage.

The foregoing re-introduces the matter of single-phase isentropic changes or pathways which diverge away from, or converge with, the vapor-pressure curve. In further comment, a divergent isentropic compression corresponds to a convergent isentropic expansion, and a convergent isentropic compression corresponds to a divergent isentropic expansion.

With regard to isentropic-type changes which occur in the two-phase region, the situation is further complicated by the latent heat effects which occur simultaneously. That is, condensation will give off heat, whereas vaporization will require heat. Furthermore, depending upon the divergent vs. convergent isentropic behavior relative to the vapor-pressure curve, either condensation of vaporization may occur, with either compression or expansion.

The foregoing considerations lead to a further examination of energy cycles in which the working fluid remains at a condition of saturation or near-saturation—that is, the working fluid is maintained as a saturated vapor or nearly so. Thus energy cycles in which there is featured the compression of a saturated-vapor along or near the vapor-pressure curve or saturated-vapor locus can, in principle, have certain advantages in efficiency. These advantages, such as they are, will be examined more fully in Chapter 7 for power generation and in Chapter 8 for heat pump cycles.

To be dispensed with first, there is the matter of phase change and pressure-volume work. That is, under what circumstances if any can changes in phase, with accompanying latent heat effects, be converted to pressure-volume work—or vice versa? This is followed by a detailed consideration of the nuances involved and the determination of pressure-volume work requirements when the working fluid follows the vapor-pressure or saturation curve.

6.1. LATENT HEAT CHANGE AND WORK

The enthalpic latent heat change (ΔH) at constant pressure, which pertains to an open or flow system, does not result in the exchange of pressure-volume work; that is, in the differential form, $V\,dP = 0$, although the volume or specific volume V varies. In another manner of speaking, *isobaric* enthalpic latent heat changes cannot convert directly, that is, unaided, to pressure-volume work changes or vice versa.

An illustration is that a saturated vapor passing through an adiabatic turbine will not result in condensation of the vapor to produce movement of the turbine if the exact same constant pressure exists on both sides of the turbine. Similarly, in an adiabatic closed system, a saturated vapor will not condense, unaided, to produce movement of a confining piston—assuming the exact same forces or pressures exist on both ends of the piston. That is, the presure stays constant at equilibrium.

The argument may be extended to nonadiabatic systems. Thus consider a closed system that consists of a saturated liquid–vapor mixture in a cylinder with a (frictionless) confining piston. The addition or removal of heat will cause movement of the piston as vaporization or condensation occurs. For the system to remain at the exact same constant pressure, however, the movement of the piston must be counterbalanced exactly by an identical force or pressure that acts in the opposite direction. Thus the work term in one direction will be counteracted identically by a work term in the opposing direction. Furthermore, because there is no change in pressure, there is no movement of the equivalent dead weight, which is the basic measure of pressure, as per Section 1.1.

It can be concluded, therefore, that the differential form $-P\,dV$ as the measure of work for a closed system requires that P vary (as V varies), and for an open or flow system the differential form $V\,dP$ also requires that the pressure vary (as the specific volume V varies). Thus, strictly speaking, pressure-volume work cannot occur in an absolutely incompressible fluid.

Adiabatic Pressure-Drop

If there is an adiabatic reduction in pressure, however, a pressure–volume work exchange can occur. This is, in fact, the function of the two-phase turbine: to extract enthalpic energy from a two-phase expansion.

Nonadiabatic Condensation (or Vaporization)

Alternately, heat can be removed from (or added to) the system to cause condensation (or evaporation or vaporization). This heat can, therefore, result in a work exchange. Thus there can be a heat–work exchange across or along the saturation curve, but at nonadiabatic conditions.

It can be argued, however, that the condensation or evaporation produces a pressure difference. Therefore, the case again becomes that of two-phase expansion (or compression) under a pressure difference.

Open or Flow Systems versus Closed Systems

In the final analysis, enthalpic latent heat change is defined or constructed whereby no pressure–volume work exchange can occur. That in effect is what is meant by the enthalpic latent heat change: it is to be purely thermal. Thus if the enthalpic latent heat exchange, say $\Delta H = \Delta H_v$, applies for an open or flow system, then for a closed system the latent internal energy change $\Delta U = \Delta U_v$ is related by

$$\Delta U_v = \Delta U_v + \Delta(PV)$$
$$= \Delta U_v + P\Delta V + V\Delta P$$

where the subscript v is here used to denote vaporization. Since $V \Delta P = 0$, then in consistent units,

$$\Delta U_v = \Delta H_v - P \Delta V,$$

where, in this case, $\Delta V = V_V - V_L$ would be the difference between the specific or molar volumes of the vapor and liquid phases.

If V_V is controlling such that $V_V - V_L \sim RT/P$, then

$$\Delta U_v = \Delta H_v - RT$$

or

$$\frac{\Delta U_v}{T} = \frac{\Delta H_v}{T} - R.$$

The connotations are similar to $C_v = C_p - R$.

Ordinarily, ΔU_v would not be measurable directly unless the vaporization (or condensation) could be conducted in a cell at a constant pressure. This situation would require an expanding (or contracting) barrier that could be maintained at a constant internal pressure, for example, an ideal membrane or (frictionless) piston.

Note that with regard to the Clausius–Clapeyron relationship

$$\frac{dP}{dT} = \frac{\Delta H_v}{T(V_V - V_L)},$$

substitution for ΔH_v yields

$$\frac{dP}{dT} = \frac{\Delta U_v + P(V_V - V_L)}{T(V_V - V_L)}$$

$$= \frac{\Delta U_v}{T(V_V - V_L)} + \frac{P}{T}$$

or

$$\frac{d \ln P}{d \ln T} = \frac{\Delta U_v}{P(V_V - V_L)} + 1$$

$$\sim \frac{\Delta U_v}{RT} + 1$$

or

$$\frac{d \ln P}{dT} \sim \frac{\Delta U_v}{RT^2} + \frac{1}{T}.$$

Therefore,

$$\frac{\Delta H_v}{RT^2} \sim \frac{\Delta U_v}{RT^2} + \frac{1}{T}$$

or

$$\Delta U_v \sim \Delta H_v - RT.$$

This can be used to estimate ΔU_v, and is consistent with the previous result.

6.2. TWO-PHASE EXPANSION AND COMPRESSION

The expressions for isenthalpic and isentropic single-phase expansions are reviewed as follows, particularly as are applied to two-phase expansion and compression.

Isenthalpic Gaseous Expansion

In an adiabatic isenthalpic expansion of a gas, no work is done. Under these conditions, the Joule–Thomson coefficient applies:

$$\frac{dT}{dP} = \mu,$$

where μ represents the Joule–Thomson coefficient, which is the change in temperature with respect to pressure, denoted here as dT/dP or as $\Delta T/\Delta P$, where Δ represents an incremental change or difference. Experimental values of μ are generally available for a number of common gases as a function of temperature and pressure, and can be estimated for other gases. In the domain of interest, μ is most generally positive, although it may be negative in other regions of temperature and pressure. The crossover where $\mu = 0$ is called the inversion curve. The subject is further discussed, for instance, in reference 1, including the role of the Joule–Thomson coefficient in the adjustment of the thermodynamic temperature scale.

For the purposes of estimation,

$$-C_p \mu = V - T \left(\frac{\partial V}{\partial T} \right)_P,$$

where the partial derivative is determined from the equation of state for the particular gas under consideration.

Isentropic Gaseous Expansion

For the isentropic expansion of a gas, the relationships are of the same form as for compression. Thus in terms of the symbols used, for a flow system, for a gaseous or vapor phase, in the differential form and in consistent units, per unit mass of mole of gas,

$$dQ - dW = C_p\, dT = V\, dP \quad \text{or} \quad C_p d \ln T = R d \ln P,$$

and since the expansion is also adiabatic,

$$-dW = C_p\, dT$$

The first equation integrates to the usual form for an isentropic compression or expansion. In an expansion, dT will be negative and dW will be positive, which indicates work done by the system. The theoretical work and actual work delivered during expansion are related by the efficiency, which may be on the order of 80%, plus or minus.

Although it is not necessary to the success of heat pumps, say, the use of a mechanical expander or turbine can add flexibility. The log mean temperature differences at the heat exchangers can be increased, which causes a reduction in exchanger area. The choice becomes a trade-off between expander efficiency and cost, and heat exchanger size and cost. An additional complication to be addressed is that the mass or moles of gas V changes during a two-phase expansion or compression.

Isenthalpic Two-Phase Expansion

The previously stated relationships pertain to the single-phase expansion of a gas, either isenthalpically or isentropically. If the expansion involves the phase change of a liquid phase or condensate into a gas, then other relationships apply.

The derivation of these relationships is as follows (2): for the adiabatic case where no work is done (isenthalpic expansion),

$$F = L + V \quad \text{(material balance)},$$

$$FH_F = LH_L + VH_V \quad \text{(enthalpy balance)},$$

where

$$F = \text{flow rate of feed to expander},$$

$$L = \text{flow rate of exiting liquid phase, if any},$$

$$V = \text{flow rate of exiting vapor phase},$$

H_F = enthalpy of feed stream,

H_L = enthalpy of exiting liquid phase, if any,

H_V = enthalpy of exiting vapor phase.

The operation is diagrammed schematically as follows, for expansion via a throttle valve:

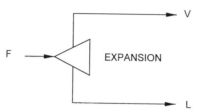

Note that the stream F also may be in two phases before the expansion. As a simplification

$$FH_F = LH_L + V(H_L + \Delta H) \quad \text{(energy balance)},$$

where $\Delta H = \Delta H_v = \lambda$ is the latent heat of vaporization. By rearranging, since

$$F(H_F - H_L) = V\Delta H,$$

if F is a liquid phase, then

$$F(C_p)_L (T_F - T_L) = V\Delta H$$

or

$$\frac{V}{F} = \frac{(C_p)_L (T_F - T_L)}{\Delta H}.$$

The difference $T_F - T_L = T_F - T_V$, is the temperature drop across the expansion or throttle valve, where $T_L = T_V$ at phase equilibrium. Furthermore, let this expansion be a leg of some energy conversion cycle 1-2-3- $\cdots$, such that say T_F corresponds to T_2 and $T_L = T_V$ corresponds to T_3, where the expansion is from pressure level P_B to pressure level P_A. Therefore,

$$\frac{V}{F} = \frac{(C_p)_L (T_2 - T_3)}{\Delta H}.$$

For complete vaporization, $V/F = 1$, so that

$$1 = \frac{(C_p)_L(T_2 - T_3)}{\Delta H}$$

or

$$T_2 - T_3 = \frac{\Delta H}{(C_p)_L}.$$

The foregoing implies that $\Delta H = \Delta H_v$ is a constant.

Isentropic Two-Phase Expansion

If a transfer of work and/or heat with the surroundings is involved, then the energy balance may be stated, in heat units, as follows:

$$Q - W = LH_L + VH_V - FH_F = LH_L + V(H_L + \Delta H) - FH_F$$
$$= F(H_L - H_F) + V\Delta H.$$

In the differential form, where H_F is a constant,

$$dQ - dW = F(C_p)_L \, dT + \Delta H \, dV,$$

where at saturation or equilibrium $T_V = T_L = T$.
   Alternately,

$$Q - W = L(H_V - \Delta H) + VH_V - FH_F = F(H_V - H_F) - L\,\Delta H$$

or

$$dQ - dW = F(C_p)_V \, dT - \Delta H \, dL.$$

Since $dL = -dV$, it is sufficient to assume that at saturation $(C_p)_L = (C_p)_V = C_p$.
   In turn, for an adiabatic change, in heat units,

$$-dW = \left[L\bar{V}_L + V\bar{V}_V \right] dP + d\Omega.$$

Therefore, for an isentropic expansion, where $d\Omega = J_0 \, d\omega = 0$,

$$FC_p \, dT + \Delta H \, dV = \left[L\bar{V}_L + V\bar{V}_V \right] dP,$$

where $F = V + L$ and F is constant (and may be set equal to unity for most purposes). Furthermore, the molar volume of a perfect gas is RT/P, whereas for the liquid phase the specific or molar volume may be assumed constant.

Substituting the gas law for the specific or molar volume of the vapor, the resulting equation is a total differential equation in three variables: temperature, pressure, and the molar vapor flow rate V. The solution, however, is subject to the vapor-pressure relationship between pressure and temperature, which will be discussed next.

Expansion and Compression along the Vapor-Pressure Curve

The preceding derivations were set up independently of any other constraint that might exist between temperature and pressure. However, if the locus of expansion or compression follows the vapor-pressure curve, then this is an a priori constraint between temperature and pressure. Two situations will be examined: the limiting case with no phase change and the more general case with phase change.

No Phase Change. It may be assumed that $dV = 0$ and

$$FC_p \, dT = \left[L\bar{V}_L + V\bar{V}_V \right] dP = L\bar{V}_L \, dP + V\frac{RT}{P} \, dP,$$

where both V and L are constant. Solution in terms of T and P would require numerical integration, starting from some point (T_1, P_1). If, however, the volume of vapor is very much greater than the volume of liquid, then

$$FC_p \, dT = V\frac{RT}{P} \, dP.$$

Integrating between limits,

$$\frac{T_2}{T_1} = \left(\frac{P_2}{P_1} \right)^{(R/C_p)(V/F)} = \left(\frac{P_2}{P_1} \right)^{k/(k-1)(V/F)},$$

which is of the same form as that for the isentropic expansion of a gas. For the energy requirement,

$$Q - W = FC_p(T_2 - T_1),$$

where the temperature difference would be determined as before and where $Q = 0$ for an adiabatic isentropic expansion.

Phase Change. If the phases are at saturation or equilibrium, and remain so, the locus of expansion will be defined by the vapor-pressure curve, whereby it may be assumed that

$$\frac{dP}{dT} = \frac{P\Delta H}{RT^2} \quad \text{or} \quad \ln\frac{P_2}{P_1} = -\frac{\Delta H}{R}\left(\frac{1}{T_2} - \frac{1}{T_1}\right).$$

Furthermore, by the energy balance, where $d\Omega = J_0\, d\omega = 0$,

$$FC_p + \Delta H\frac{dV}{dT} = \left[L\overline{V}_L + V\overline{V}_V\right]\frac{dP}{dT} = \left[L\overline{V}_L + V\frac{RT}{P}\right]\frac{P\Delta H}{RT^2}$$

or

$$\frac{dV}{dT} = \left[L\overline{V}_L + V\frac{RT}{P}\right]\frac{P}{RT^2} - \frac{FC_p}{\Delta H}.$$

Solution requires numerical integration starting from some point (T_1, P_1) on the vapor-pressure curve, where $V = V_1$ (and may include as the limiting cases $V_1 = 0$ and $V_1 = F$). The integration would continue to some other point on the curve (T_2, P_2), where V will have the value V_2. Temperature may be assumed to be the independent variable for successive increments δT. Furthermore, by virtue of the vapor-pressure curve, the increments δT will be negative for an expansion (where δP is negative) and positive for a compression (where δP is positive).

Analytic integration will suffice if it is assumed that the vapor molar rate or volume is markedly greater than the liquid molar rate or volume, and remains so. (This will exclude small values of V/F from consideration, including $V/F = 0$). Accordingly,

$$\frac{dV}{dT} = V\frac{RT}{P}\frac{P}{RT^2} - \frac{FC_p}{\Delta H} = \frac{V}{T} - \frac{FC_p}{\Delta H} = \frac{V}{T} - a,$$

where a is defined by the substitution $a = FC_p/\Delta H$. By the expediency of assuming

$$\frac{V}{T} = f \quad \text{or} \quad V = Tf \quad \text{or} \quad dV = Tf' + f,$$

where $f = f(T)$ is a function of T, then

$$Tf' + f = \frac{V}{T} - a = f - a,$$

whereby

$$T\frac{df}{dT} = -a.$$

Integrating,

$$f = \frac{V}{T} = -c \ln T + c \quad \text{or} \quad V = -aT \ln T + cT,$$

where c is the constant of integration, established from the boundary conditions, for a point on the vapor-pressure curve.

(On differentiating,

$$\frac{dV}{dT} = -a\left[T\frac{1}{T} + \ln T\right] + c = -a + (-a \ln T + c) = -a + \frac{V}{T},$$

which constitutes the original differential equation.)

As the result, with respect, say, to a point (T_1, P_1), where $V = V_1$,

$$\frac{V_2}{T_2} - \frac{V_1}{T_1} = -\frac{FC_p}{\Delta H}\ln\frac{T_2}{T_1},$$

where also, by virtue of the vapor-pressure curve,

$$\frac{P_2}{P_1} = \exp\left[-\frac{\Delta H}{R}\left(\frac{1}{T_2} - \frac{1}{T_1}\right)\right] \quad \text{or} \quad \frac{1}{T_2} - \frac{1}{T_1} = -\frac{R}{\Delta H}\ln\frac{P_2}{P_1}.$$

Thus the second point can be related to the first point. By definition, however, the second point also must lie on the vapor-pressure curve, and both V_1/F and V_2/F cannot exceed unity. (Note that although a solution would seem to occur for the case where, say, the initial vapor-phase flow rate $V_1 = 0$, this is not allowable because it negates the original assumption that the vapor volumetric rate is controlling; that is, the liquid volumetric rate can be neglected.)

For the statement about energy exchange,

$$Q - W = FC_p(T_2 - T_1) + \Delta H(V_2 - V_1),$$

where $Q = 0$ for an adiabatic expansion or compression. The values for T_2 and V_2 are obtained by the foregoing integration.

Check. As a check, for the work term,

$$-W = \int V\bar{V}\,dP = \int V\frac{RT}{P}\,dP = \int V\frac{RT}{P}\frac{dP}{dT}\,dT.$$

On substituting the vapor-pressure relation $dP/dT = P\Delta H/RT$, we obtain

$$-W = \int \frac{V}{T}\Delta H\,dT = \int \left[\frac{V_1}{T_1} - \frac{FC_p}{\Delta H}\ln\frac{T}{T_1}\right]\Delta H\,dT.$$

Integrating, between T_1 and T_2,

$$-W = \frac{V_1\,\Delta H}{T_1}(T_2 - T_1) - FC_p T_1\left[\frac{T_2}{T_1}\ln\frac{T_2}{T_1} - \frac{T_2}{T_1} + 1\right]$$

$$= \Delta H\left(\frac{V_1}{T_1} + \frac{FC_p}{\Delta H}\right)(T_2 - T_1) - \Delta H\left(\frac{FC_p}{H}\right)\ln\frac{T_2}{T_1}$$

$$= \Delta H V_1\frac{T_2}{T_1} - \Delta H V_1 + FC_p(T_2 - T_1) + \Delta H T_2\left(\frac{V_2}{T_2} - \frac{V_1}{T_1}\right)$$

$$= \Delta H(V_2 - V_1) + FC_p(T_2 - T_1).$$

The check indicates that the derivations are consistent.

Note. It may be observed that

$$\frac{d(V/F)}{dT} = \frac{V/F}{T} - \frac{C_p}{\Delta H}.$$

Whether or not V or V/F increases as T decreases will depend upon the relative size of

$$\frac{V/F}{T}\quad\text{versus}\quad\frac{C_p}{\Delta H}.$$

Starting at a point where $V/F = 1$, that is, stream V is 100% saturated vapor, if the difference $[1/T - (C_p/\Delta H)]$ is negative, then for *compression* dT is positive and $d(V/F)$ will be negative; that is, condensation will occur during compression. If the difference were positive, then $d(V/F)$ would be positive, which signifies that the path of isentropic compression would then have to be away from the vapor-pressure curve and into the gaseous region.

Starting again, at a point where $V/F = 1$, if the difference is negative, then for *expansion dT* is negative and $d(V/F)$ would be positive, which signifies that the path of expansion would have to be away from the vapor-pressure curve and into the gaseous region. If the difference were positive, however, then $d(V/F)$ would be negative and condensation could occur during an isentropic expansion.

These particular results may be generalized, starting at some intermediate point that is part vapor and part liquid. Thus in compression, if $[1/T - (C_p/\Delta H)]$ is negative, then further condensation will occur with increasing pressure and temperature. If $[1/T - (C_p/\Delta H)]$ were positive, however, further vaporization could occur with increasing temperature and pressure. In expansion, if $[1/T - (C_p/\Delta H)]$ is negative, then further vaporization will occur with decreasing temperature and pressure. If $[1/T - (C_p/\Delta H)]$ were positive, however, then further condensation could occur with decreasing temperature and pressure. At the point where $1/T = (C_p/\Delta H)$, presumably an inversion condition will exist where there is no change in the vapor–liquid ratio with either compression or expansion.

Work. Returning to the solution for V,

$$V = -aT \ln T + cT,$$

from which c may be determined as $c = (V_1/T_1) + aT_1 \ln T_1$. Substituting for a and rearranging,

$$\frac{V}{F} - \frac{V_1}{F} = -\frac{C_p}{\Delta H} T \ln \frac{T}{T_1} + \frac{V_1}{F}\left[\frac{T}{T_1} - 1\right].$$

Therefore, for the adiabatic work along the vapor-pressure curve, consider the differential form

$$-d\left(\frac{W}{F}\right) = C_p \, dT + \Delta H d\left(\frac{V}{F}\right)$$

$$= [C_p + \Delta H]\frac{d(V/F)}{dT} dT$$

$$= [C_p + \Delta h]\left[\frac{V/F}{T} - \frac{C_p}{\Delta H}\right] dT,$$

which integrates to the following expressions between temperature levels T_1 and T_2:

$$-\frac{W}{F} = C_p(T_2 - T_1) + \Delta H \left[\frac{V_2}{F} - \frac{V_1}{F} \right]$$

$$= C_p(T_2 - T_1) - C_p T_2 \ln\frac{T_2}{T_1} + \frac{V_1}{F} \Delta H \left[\frac{T_2}{T_1} - 1 \right].$$

Whether $(-W)$ is positive or negative during a compression or expansion will depend upon the temperatures and temperature range, and the values for C_p and ΔH.

Note that an inversion will occur in the differential form whenever the difference $(V/F)/T - C_p/\Delta H = 0$. Additionally, whenever this difference is positive, $(-dW)$ will be positive for compression and negative for expansion. Conversely, whenever this difference is negative, $(-dW)$ will be negative for compression and positive for expansion. The overall integrated result will depend upon the particular values for T_2 and T_1, and the values for C_p and ΔH, as previously commented. In other words, the latent heat effects may work for or against the compressor work input or the expander work done, depending upon the particular system and the temperatures involved.

Total Flow Expander. A total flow expander may be employed (2), as per Fig. 6.1. The work W done by the expander is related mostly to the expansion of the gas. The gas-phase rate may remain constant or may vary, as previously noted where the vapor–liquid system is at saturation. In any event, the work produced will be less than for the isentropic expansion where the total stream is a gas.

Compression versus Vapor Pressure

For logarithmic coordinates, the compression curve may be transformed as follows:

$$\frac{T}{T_4} = \left(\frac{P}{P_4} \right)^{(k-1)/k} \quad \text{or} \quad \left(\frac{T}{T_4} \right)^{k/(k-1)} = \frac{P}{P_4}.$$

Taking the logarithm,

$$\ln\frac{P}{P_4} = \frac{k}{k-1}\ln\frac{T}{T_4}.$$

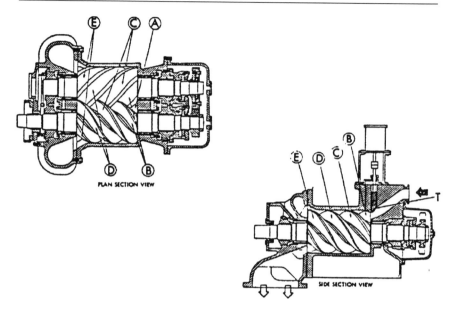

Fig. 6.1. Helical rotary screw expander. (Hydrothermal Power Co., Ltd. Consult R. A. McKay, "The Helical Expander Evaluation Project," Twelfth Intersociety Energy Conversion Engineering Conference, Washington, DC, August–September, 1977.)

This will be a straight line on a log-log plot, starting at, say, a point (P_4, T_4). The log-log slope will be

$$\text{slope of compression locus} = \frac{k}{k-1}.$$

Over restricted ranges of temperature and pressure, the vapor-pressure curve also will appear as a straight line on a log-log plot, as shown at the end of Section 2.2; that is

$$\frac{P}{P_a} = \left(\frac{T}{T_a}\right)^m$$

or

$$\ln\frac{P}{P_a} = m\ln\frac{T}{T_a},$$

where m is the slope and (P_a, T_a) is some reference point on the vapor-pressure curve.

In comparing the compression locus versus the vapor-pressure curve, the compression locus may converge toward or diverge away from the vapor-pressure curve. These situations are indicated schematically in Fig. 6.2, starting from a common point, indicated as (P_4, T_4). In Fig. 6.2(a) the representation is in P-T coordinates; in Fig. 6.2(b) the representation is in log P-log T coordinates. The converging compression locus in each case is shown extended beyond the vapor-pressure curve, whereas the locus would actually stop at the vapor-pressure curve and would then follow the curve upward for so-called wet compression, or else the condensate or liquid phase could, in principle, be removed as it is formed, which leaves only the saturated vapor phase to be compressed.

This means that in a compression–expansion cycle, in the converging case, the inlet vapor must be superheated to a sufficient degree to remain outside the vapor-pressure curve during compression. In the diverging case, the exit vapor will become excessively superheated, which represents an excessive compressor work requirement. The two situations are compared in Figs. 6.3 and 6.4 in P-T coordinates.

The ideal situation is where the compression locus parallels the vapor-pressure locus. For this situation the compressor inlet and exit vapors will involve only a low degree of superheat. In other words, the compression locus may be virtually congruent with the vapor-pressure curve.

For Freon-21, for instance, the compression locus and vapor-pressure curve have virtually the same slope, whereas for water vapor the compression locus diverges from the vapor-pressure curve, which produces a superheating of the compressor outlet vapors.

Freon-21 (Dichloromonofluoromethane)

Vapor pressure	Temperature		
10 atm	87°C	185°F	645°R
20 atm	121°C	250°F	710°R

$$\ln\frac{20}{10} = \frac{m}{n}\frac{710}{645}, \qquad m = 7.22,$$

$$k = 1.157, \qquad \frac{k}{k-1} = 7.1.$$

The log-log slopes are virtually equal.

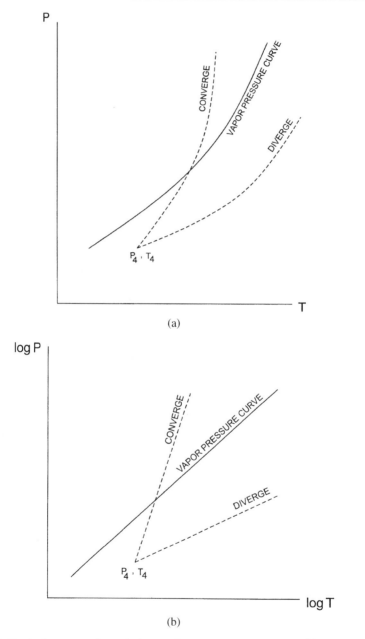

Fig. 6.2. Loci of compression relative to the vapor-pressure curve. (a) *P-T* coordinates; (b) log *P*-log *T* coordinates.

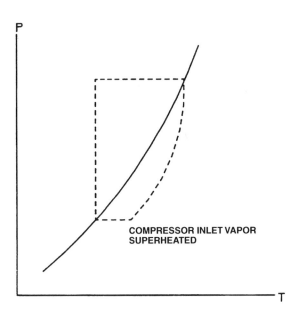

Fig. 6.3. Inlet vapor superheated.

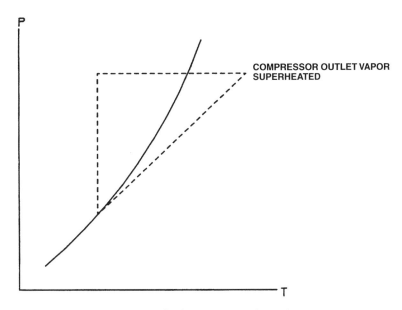

Fig. 6.4. Outlet vapor superheated.

Water

Vapor pressure	Temperature	
11.5 psi	200°F	660°R
49.2 psi	280°F	740°R

$$\ln\frac{49.2}{11.5} = m, \qquad \ln\left(\frac{740}{660}\right), \qquad m = 12.70,$$

$$k = 1.316, \qquad \frac{k}{k-1} = 4.17.$$

The log-log slope of the compression locus is much less than the slope of the vapor pressure curve.

The convergence or divergence can be used to estimate the degree of superheat at the compressor inlet or outlet. Thus consider the pressure range between P_A and P_B, whereby

$$\frac{T_{\text{comp}}}{T_a} = \left(\frac{P_B}{P_A}\right)^{(k-1)/k} \quad \text{and} \quad \frac{T_{\text{vp}}}{T_a} = \left(\frac{P_B}{P_A}\right)^{1/m}$$

and where the two curves intersect at P_A and T_a. The temperature T_{comp} is the temperature along the locus of compression at $P = P_B$. The temperature T_{vp} is the corresponding temperature on the vapor-pressure curve at $P = P_B$. The degree of superheat at pressure P_B is given by the difference $T_{\text{comp}} - T_{\text{vp}}$:

$$T_{\text{comp}} - T_{\text{vp}} = T_a\left[\left(\frac{P_B}{P_A}\right)^{(k-1)/k} - \left(\frac{P_B}{P_A}\right)^{1/m}\right].$$

This relationship applies to either convergence or divergence, and in either direction.

The foregoing calculations also may be adapted to expansion as well as compression.

Two-Phase Region

Ordinarily a single component will be used as the working fluid. In this context it is proper to denote the two-phase (vapor–liquid) region by use of the vapor-pressure curve. However, mixtures of two or more components may be used (multicomponent mixtures). In this case, the two-phase

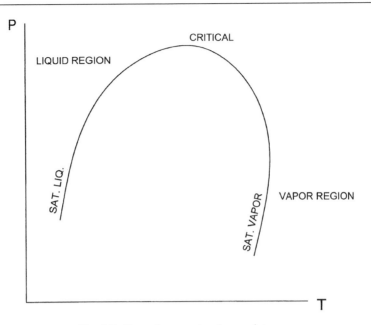

Fig. 6.5. Two-phase region for a mixture.

region will exist as an envelope, as illustrated schematically in Fig. 6.5. The features are as follows.

The outer loci denote the saturation curve of 100% liquid and 100% vapor, respectively. In the interior, different curves may be used to represent a constant percentage or fraction of liquid or vapor. All the curves will terminate at the critical point—the point where liquid and vapor become indistinguishable and where the latent heat of vaporization approaches zero. Beyond the critical point is the so-called supercritical region, the single-phase region where what are thought of as gases appear more as liquids or vice versa. The supercritical region is a transition region between what is considered the gaseous state and what is considered the liquid state. For a single-component system, the saturated liquid and vapor lines in effect merge to form a single curve—the vapor-pressure or saturation curve—that culminates in the critical point.

More properly, the terms "vapor" and "liquid" are distinguishable only when they are in equilibrium (also called saturation). Away from the equilibrium condition, the vapor more properly is referred to as a gas or as a superheated vapor or gaseous phase. Away from the equilibrium condition, the liquid more properly is referred to as a supercooled liquid.

Inasmuch as the descriptions and derivations will apply to a mixture as well as a single component, the term "vapor-pressure curve" will be intended to apply to a mixture as well as to a single component. The term saturation curve also may be applied; it denotes that the phases exist at phase equilibrium.

It will be understood, furthermore, that under the appropriate conditions, compression (or expansion) can occur along or near the saturated vapor locus (100% vapor line in the two-phase envelope, also called the dew-point curve). Such a compression (or expansion) locus will require the injection or removal of liquid. These special circumstances will be further examined.

Liquid Injection or Removal at Saturation

Compression (or expansion) may be conducted along the saturated vapor curve utilizing latent heat effects to retain the vapor at saturation. Thus a (saturated) liquid phase may be injected or withdrawn. The differential analysis is somewhat similar to that for adiabatic two-phase behavior with phase change except that the vapor flow rate is not subject to a total mass balance where the total flow rate remains constant. Rather the vapor flow rate is affected only by the liquid vaporized or condensed.

The case where an isentropic compression of the vapor or gaseous phase would tend to diverge away from the vapor-pressure curve is diagrammed in Fig. 6.6. The object here is for the path of compression (or expansion) to lie instead along the vapor-pressure curve.

The differential energy balance may be written in heat units as

$$dQ - dW = VC_p \, dT + \Delta H \, dV = V\overline{V} \, dP + d\Omega,$$

where for an adiabatic isentropic change, $dQ = 0$ and $d\Omega = J_0 \, d\omega = 0$. The term $VC_p \, dT$ replaces the term $FC_p \, dT$ previously used, where F was the total flow rate and a constant. In other words, a separate and distinct saturated liquid phase does not exist. Liquid, if injected, is vaporized; if condensed from the vapor, the liquid is removed. Moreover, the vapor phase V will be a variable as the operation proceeds. Since $V < F$, the work exchange obviously will be less, other things being equal.

For a perfect gas, therefore,

$$VC_p + \Delta H \frac{dV}{dT} = V \frac{RT}{P} \frac{dP}{dT}.$$

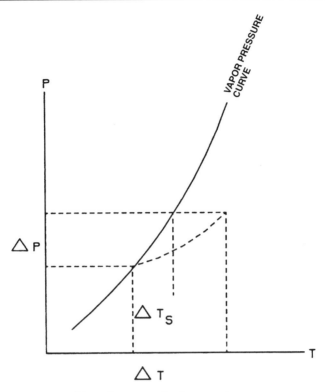

Fig. 6.6. Juxtaposition of compression locus and vapor-pressure curve.

The path of compression (or expansion) will lie along the vapor-pressure curve. Substituting the vapor-pressure relation $dP/dT = P\,\Delta H/RT$ and rearranging yields

$$\frac{d\ln V}{dT} = \frac{1}{T} - \frac{C_p}{\Delta H}.$$

Integrating with respect to a reference condition designated as point 1 yields

$$\ln\frac{V}{V_1} = \ln\frac{T}{T_1} - \left(\frac{C_p}{\Delta H}\right)(T - T_1)$$

or

$$\frac{V}{T} = \frac{V_1}{T_1}\exp\left[-\frac{C_p}{\Delta H}(T - T_1)\right].$$

The preceding equation relates V and T. Pressure can be introduced via the integrated vapor-pressure relationship

$$\ln\frac{P}{P_1} = -\frac{\Delta H}{R}\left(\frac{1}{T} - \frac{1}{T_1}\right) \quad \text{or} \quad \frac{P}{P_1} = \exp\left[-\frac{\Delta H}{R}\left(\frac{1}{T} - \frac{1}{T_1}\right)\right].$$

In sum, a way is provided to relate V and T, P and T, or P and V, relative to some reference condition.

The work of adiabatic compression is given by

$$-W = \int_{T_1}^{T} VC_p\, dT + \Delta H(V - V_1).$$

The integral can be evaluated by parts:

$$\int VC_p\, dT = C_p\frac{V_1}{T_1}\int T\exp\left[-\frac{C_p}{\Delta H}(T - T_1)\right] dT = C_p\frac{V_1}{T_1}(I - I_0),$$

where

$$I - I_0 = T\left(\frac{-C_p}{\Delta H}\right)^{-1}\exp\left[-\frac{C_p}{\Delta H}(T - T_1)\right]\bigg| - \left(-\frac{C_p}{\Delta H}\right)^{-1}$$

$$\times \int \exp\left[-\frac{C_p}{\Delta H}(T - T_1)\right] dT$$

$$= \left(-\frac{C_p}{\Delta H}\right)^{-1}\exp\left[-\frac{C_p}{\Delta H}(T - T_1)\right]\left[T - \left(-\frac{C_p}{\Delta H}\right)^{-1}\right]\bigg|_{T_1}^{T}.$$

Thus

$$-W = C_p\frac{V_1}{T_1}(I - I_0) + \Delta H(V - V_1).$$

Alternately and more simply, on introducing the vapor-pressure relation,

$$-W = \int V\bar{V}\, dP = \int V\frac{RT}{P}\, dP = \Delta H\int\frac{V}{T}\, dT$$

$$= \Delta H\frac{V_1}{T_1}\int \exp\left[-\frac{C_p}{\Delta H}(T - T_1)\right] dT$$

$$= \Delta H\frac{V_1}{T_1}\left(-\frac{C_p}{\Delta H}\right)^{-1}\exp\left[-\frac{C_p}{\Delta H}(T - T_1)\right]\bigg|_{T_1}^{T}.$$

The one form will transform to the other. The calculation is sensitive to $C_p/\Delta H$.

Note. The vapor phase V will increase during compression due to the injection of liquid, which is vaporized (and uses up latent heat), and since V increases during the compression from some initial value V_1 to some larger final value V_2, then the work of compression $(-W)$,

$$-W = \int V \bar{V} dP < \int V_2 \bar{V} dP,$$

will be less than the corresponding work of compression for an isentropic compression for the same pressure increase or compression ratio and for the same final flow rate.

The expansion of a saturated vapor phase that exhibits convergent isentropic behavior will require the removal of liquid condensate as expansion proceeds in the two-phase region. The analysis is inversely similar, with the vapor phase V decreasing during the expansion. With divergent behavior, the expansion would be away from the vapor-pressure or saturation curve and into the gaseous region.

REFERENCES

1. Hoffman, E. J., *Analytic Thermodynamics: Origins, Methods, Limits, and Validity*, Taylor and Francis, New York, 1991.
2. Hoffman, E. J., *Phase and Flow Behavior in Petroleum Production*, Energon, Laramie WY, 1981.

CHAPTER 7

Power Cycles That
Use the Saturation Curve

The use of saturated-vapor or wet compression in modified Joule power cycles and heat pump or refrigeration cycles has the potential to increase the efficiency due to the fact that the working fluid follows the vapor-pressure curve during compression. The best of all worlds is if the fluid remains as a saturated vapor. That is, it is preferable that the working fluid consist of 100% saturated vapor. Stated differently, the working fluid stays as a vapor at its dew point.

By following the vapor-pressure curve, the final compression product is at a lower temperature than would occur for divergent isentropic compression. (The latter denotes the case where the isentropic curve diverges away from the vapor-pressure curve.) Being at a lower final temperature signals the fact that the work of compression is less than that for divergent behavior. Moreover, additional superheat will be necessary to reach the same initial temperature for the expansion, which signifies that there will be a greater difference between the expansion work done and the compressor work required, when the same level of heat rejection is assumed. The overall effect is to raise the efficiency for the conversion of heat to work. Furthermore, if the compressor gases stay at 100% saturated vapor—at the dew point—then there is no condensate liquid phase to be concerned about. Moreover, there is no loss in volume for the gaseous or vapor phase, which would in turn lower the work produced during expansion.

For convergent behavior, where the locus of compression intercepts the vapor-pressure curve, the liquid condensate can be removed and reheated as vapor for expansion, as in the Rankine cycle. The compressor requirement would decrease as the compressor vapor rate decreases. Unfortunately, expansion is away from the vapor-pressure curve, thus lowering efficiency.

For divergent behavior, where the locus of compression moves away from the vapor-pressure curve, heat-removal can be used or condensate liquids can be added, causing compression to move to the vapor-pressure curve.

Thus modified Joule power cycles with one leg along the two-phase vapor-pressure locus can afford somewhat greater efficiencies. Of particular note is the steam-water system, which exhibits divergent behavior during compression, but which can be adapted to existing steam-power plant equipment, e.g., for superheating and expansion. Such a cycle, which here can be referred to as the saturated vapor cycle for lack of a ready word—or can be called something else—is described along with the derivations and determinations for efficiency. Two main embodiments are pursued: one involves nonadiabatic compression; the other involves the injection of liquids during adiabatic compression.

In principle, vapor compression along the vapor-pressure curve can be treated nonadiabatically as described in Section 2.2 or adiabatically (isentropically) as described in Section 6.2. Inasmuch as the continued presence of a liquid phase is discouraged during compression, other expediencies are called for. These expediencies, as indicated, will involve either heat rejection during compression or the injection of condensate liquids during compression. The latter embodiment is for the purposes of following the vapor-pressure curve for the case where the working fluid exhibits divergent isentropic behavior during compression.

For purposes of comparison and reference, conventional power cycles will be reviewed first. Saturated vapor cycles for power generation, the main subject of this chapter, will be examined next, with examples and comparisons provided. Chapter 8 will examine the case for heat pump cycles.

7.1. CONVENTIONAL POWER CYCLES

The classic method used for power generation is the Rankine cycle, here illustrated again as a pressure–temperature diagram shown schematically in Fig. 7.1. Saturated steam, say, represented by point 1 is superheated to point 2 and then expanded through a turbine or turbines to point 3. Work is produced during the expansion. The exhaust gases (point 3) are cooled and then condensed at point 4. The condensed liquid or condensates is then pumped to boiler pressure, which is represented by point 5. The liquid is heated and vaporized, and returns to the saturated vapor condition as represented by point 1.

The expansion 2 to 3, moreover, may continue to or into the two-phase region, so that point 3 lies on the vapor-pressure curve. Furthermore, it is preferable that the condensate not be supercooled, so that point 4 also will lie on the vapor-pressure curve.

Working fluids other than water or steam may be utilized. The Rankine cycle is ideally suited, however, to the properties of the water-steam system. Moreover, the ubiquity of water is a great asset; that is, the low cost and ready availability.

By comparison, in the closed Joule cycle, there is no change in phase for the working fluid, which remains in the gaseous single-phase region. For reference, a schematic of the Joule cycle is diagrammed again in Fig. 7.2,

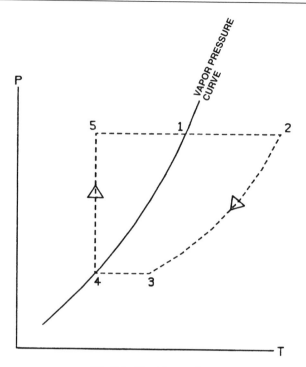

Fig. 7.1. Rankine cycle.

with the following representation. The gaseous working fluid, say, at point 1 is heated at constant pressure to point 2, and then expanded through a turbine or turbines to point 3 to produce work. The exhaust gases are cooled isobarically to point 4 and then recompressed to point 1, which completes the cycle.

The closed Joule cycle also may be replaced by an open-cycle version, usually called the Brayton cycle. This is the cycle of the gas-fired turbine, whereby hot combustion gases are used to drive a low-pressure turbine. The exhaust gases may be used to heat the combustion air, etc., to improve the efficiency.

The closed Joule cycle may, in principle, yield theoretically high efficiencies, depending upon the working fluid and the operating temperatures and pressures. Similarly, the open Brayton cycle will produce at least respectable efficiencies, particularly with waste-heat regeneration or recuperation. A characteristic of the closed Joule cycle is that the efficiency generally increases with the compression–expansion ratio, other things being equal, whereas in the open Brayton cycle, the efficiency increases

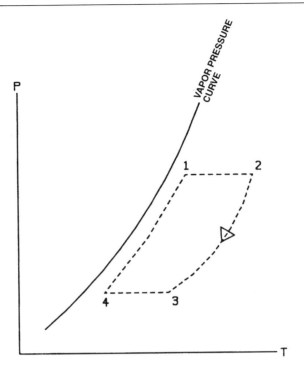

Fig. 7.2. Joule cycle.

with a decreasing compression–expansion ratio, for example, as presented in reference 1.

The afore-mentioned energy or power cycles are sometimes referred to as heat engines. The basic mathematical relationships were derived previously in Chapters 2 and 3.

7.2. THE SATURATED VAPOR CYCLE

Here what will, for convenience, be called the "saturated vapor cycle" is based on the adiabatic isentropic expansion of a gaseous superheated working fluid down to or near the saturation or vapor-pressure locus. If the locus is not quite reached, cooling may be used. Furthermore, partial condensation may occur, which is, in turn, followed by vapor recompression along the vapor-pressure curve. If the expanded gases are 100% saturated vapor, then the gases may be compressed with heat removal or rejection. The compression is therefore nonadiabatic.

If the expanded gases are partially condensed, then the condensate may be reinjected during compression and will serve to keep the path of compression along the vapor-pressure or saturation locus. The compression is then viewed as adiabatic and isentropic.

Another way of looking at it is that the so-called saturated vapor cycle is a Joule cycle where the compression leg is made to follow the vapor-pressure curve.

The idealized saturated vapor cycle is illustrated schematically in Fig. 7.3. The compressed saturated vapor represented by point 1 is superheated to point 2. The superheated vapor is expanded through a turbine or turbines to point 3 to produce work. The exhaust gases at point 3 may be cooled to the dew point to produce a 100% saturated vapor at point 4. Alternately, part of the saturated vapor may be condensed.

The expansion (points 2 to 3) may in fact continue to or into the two-phase region so that point 3 will lie on the vapor-pressure curve and will thus coincide with point 4 on the *P-T* diagram. Point 3 or point 4 can

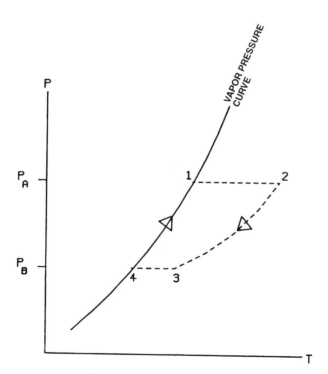

Fig. 7.3. Saturated vapor cycle.

thus represent 100% saturated vapor at the dew point or else a two-phase mixture at saturation.

The saturated vapor phase so produced is then compressed either nonadiabatically with heat rejection or else by adding sufficient liquid condensate phase to ensure that the path of compression occurs along or near the vapor-pressure curve or vapor–liquid saturation locus. Compression to point 1 completes the cycle.

The effect of heat rejection or the volatilization of injected liquid is that of offsetting the rise in temperature during compression; that is, heat rejection or condensate volatilization serves to cool the gas during compression such that the path of compression lies on, along, or near the vapor-pressure or saturation curve.

Compression may occur in several stages, with intercooling between stages or with the condensate injected before and/or after each stage or injected directly into each compressor stage. Additionally, or in lieu of the working fluid condensate, makeup working fluid liquid or liquids can be injected.

Ensuring that compression occurs along or near the vapor-pressure curve is the feature that gives the saturated vapor cycle a pronounced advantage in theoretical efficiency. As previously noted, such a compression leg may be referred to as "wet compression," although the usual inference of this term is for the case where the converging path of adiabatic compression tends to intercept the vapor-pressure curve rather than move away from the curve, as in the case at hand. Moreover, although other working fluids can be used, the saturated vapor cycle is adaptable to the considerable advantages of the steam–water system, whose locus of expansion will intercept the vapor-pressure curve.

Binary or Multicomponent Working Fluids

Instead of a single-component working fluid, binaries or multicomponent mixtures may be used, for one reason or another, such as to secure more advantageous operating conditions. In this instance, the vapor-pressure curve would be replaced by an envelope. The path of compression would tend to lie along the edge of the envelope, at 100% saturated vapor, which also is known as the dew-point curve.

If partial condensation occurs, it would be at varying temperature (at constant pressure) and toward the interior of the phase envelope. Moreover, the composition of the vapor phase and liquid or condensate phase will change as condensation proceeds. Similar circumstances exist for vaporization at a saturated or two-phase equilibrium condition. Otherwise, the same remarks apply as for single-component working fluids.

Fuel Efficiency and Operating Temperature Levels

A disadvantage if the steam–water system is used as the working fluid is the fuel efficiency. The same sort of problem exists for other cycles. The exit or stack-gas temperature of the combustive support will correspond to and be higher than T_1 in Fig. 7.3, and will ordinarily be well above ambient conditions. By using a refrigerant-type working fluid, however, the operating temperature levels can be extended downward so that T_1 corresponds more closely to ambient conditions. At the same time it will be required that $T_3 = T_4$ if there is to be no heat rejection at subambient temperatures.

Furthermore, heat rejection during compression cannot be used at subambient temperatures. Instead, the alternative of producing a condensate phase that can be injected and vaporized during compression must be used. This, however, also would call for the rejection of the latent heat of condensation at subambient temperatures. It is a stand-off.

As a final note, the leg for heat addition (1 to 2) also can be conducted at subambient temperatures. Therefore, in principle at least, the ambient surroundings could provide the heat source to, say, propel the cycle for the version of the cycle that produces condensate for revaporization during compression. Again, however, there is the problem of rejecting the latent heat of condensation at subambient conditions. The use of a heat pump, which offsets the purpose, would be required.

7.3. CYCLE EFFICIENCY

The efficiency for the cycle proper can be based on the net energy produced, that is, the work of expansion plus the work of compression (which will have opposite signs). The efficiency may be expressed either in terms of enthalpy differences or in terms of thermal properties and temperature differences. Thus for the theoretical efficiency of, say, the saturated vapor cycle, on a unit mass or molar basis,

$$\text{Eff}_{\text{SatVap}} = \frac{W_{\text{exp}} + W_{\text{comp}}}{(H_2 - H_1) + \text{EXCON}},$$

where

$$H_2 - H_1 = C_p(T_2 - T_1)$$

and where EXCON is included to represent the sensible heat change of any required condensate produced. EXCON denotes the heat required to raise the condensate from the condensation temperature to the temperature at the point of injection. This requirement will vary on a unit basis as the compression temperature increases, but a mean or average value can be used. If there is an excess of condensate beyond this requirement for vaporization during compression, then the excess condensate is assumed to be raised in temperature to T_1 and the latent heat of vaporization is added to produce the saturated vapor condition at T_1. If no condensate is produced, then EXCON = 0; that is, only heat rejection by cooling occurs during compression.

Additionally, electro-mechanical and/or mechanico-electrical conversion efficiencies can be introduced for compression and expansion. This addition will depend in part upon whether or not a direct coupling is used between the expander and compressor.

By comparison, the corresponding efficiency of the Rankine cycle in the terminology of Fig. 7.1 will be as follows, assuming the expanded vapor reaches the saturation curve:

$$\text{Eff}_{\text{Rankine}} = \frac{W_{\text{exp}}}{H_2 - (H_3)_L},$$

where $(H_3)_L$ represents the condensate enthalpy, $H_2 - (H_3)_L = (H_2 - H_3) + \lambda$, and $H_3 - H_4$ or $T_3 = T_4$. Therefore,

$$\text{Eff}_{\text{Rankine}} = 1 / \left(1 + \frac{\lambda/C_p}{T_2 - T_3} \right),$$

where $W_{\text{exp}} = C_p(T_2 - T_3)$.

The ideal Carnot efficiency could be written as

$$\text{Eff}_{\text{Carnot}} = \frac{T_2 - T_3}{T_2},$$

where for the purposes here, $T_3 = T_4$. There is not, however, a direct analogy to the Carnot cycle, or vice versa, but a reference is nevertheless provided.

Work of Compression

The determination of cycle efficiency revolves around the establishment of the work of compression. In terms of pressure–volume work,

$$-W_{\text{comp}} = \int V\bar{V} \, dP = \int V \frac{RT}{zP} \, dP = \int \Delta H \frac{V}{T} \, dT.$$

The first integral on the right, assuming a constant (mass or molar) vapor rate V, can be evaluated from a tabulation or equation of state for $\bar{V}$ versus P *at saturation*. For the steam–water system, the steam tables are the convenient reference.

The second integral, at a constant vapor rate V, could be evaluated from a vapor-pressure curve of P versus T. In turn the compressibility factor z is a function of pressure (or temperature), or it may be assumed unity for a perfect gas. The procedure is entirely equivalent to determining the utilizing the equation of state for a saturated vapor phase.

The third and last integral on the right is derived by substitution of the vapor-pressure relation $dP = P\Delta H/RT^2\,dT$. Integration can be at either a constant or varying vapor rate V. At constant V, as per Section 2.2,

$$-W_{\text{comp}} = V\Delta H \ln\frac{T}{T_1}$$

with reference to some point 1. At varying V, as per Section 6.2,

$$\frac{V}{T} = \frac{V_1}{T_1}\exp\left[\frac{-C_p}{\Delta H}(T - T_1)\right]$$

and

$$-W_{\text{comp}} = \Delta H\frac{V_1}{T_1}\left(\frac{-C_p}{\Delta H}\right)^{-1}\left[\exp\left[\frac{-C_p}{\Delta H}(T - T_1)\right] - 1\right].$$

The foregoing equation will be utilized in an example problem.

EXAMPLE 7.1

Superheated steam is to be expanded to the vapor-pressure curve at 120°F and 1.7 psia. This will denote the exhaust conditions from the expander. The initial pressure of the superheated steam is to be 425 psia. The following physical properties are assumed:

$$C_p = 0.5 \text{ Btu/lb-°F}, \qquad \lambda = \Delta H = 900 \text{ Btu/lb}, \qquad \frac{k}{k-1} = 0.22 \text{ to } 0.23.$$

Accordingly, for the expansion, as per Fig. 7.3 where $T_3 = T_4$,

$$\frac{T_2}{T_3} = \left(\frac{425}{1.7}\right)^{0.23} = 3.5606$$

and

$$T_2 = (120 + 460)(3.5606) = 2065°R \text{ or } 1605°F \text{ rounded off to } 1600°F.$$

If the steam tables are consulted, the entropy at point 3 is 1.9339 Btu/°R and at point 2 in 1.9381. If the expansion were isentropic according to the steam tables, then for an entropy of 1.9339 at 425 psia, the corresponding temperature would be about 1560°F.

For the heat added along leg 1-2,

$$H_2 - H_1 = C_p(T_2 - T_1) = 0.5(1600 - 450) = 575 \text{ Btu/lb},$$

where 450°F is approximately the saturation temperature at 425 psia.

For the work of expansion along leg 2-3,

$$W_{\text{exp}} = 0.5(1600 - 120) = 740 \text{ Btu/lb}.$$

Alternately the steam tables could be consulted.

For the work of compression at a constant vapor rate V with heat rejection,

$$-W_{\text{comp}} = 900 \ln \frac{910}{580} = 405 \text{ Btu/lb},$$

where

$$T_1 = 450 + 460 = 910°R, \qquad T_3 = T_4 = 120 + 460 = 580°R.$$

Therefore, for the theoretical efficiency with compressor heat rejection,

$$\text{Eff}_{\text{SatVap}} = \frac{740 - 405}{575} = 58.3\%.$$

For the work of compression with partial condensation and revaporization, assuming the final vapor rate to be unity at point 1,

$$\frac{1}{910} = \frac{V_4}{580} \exp\left[\frac{-0.5}{900}(910 - 580)\right],$$

from which $V_4 = 0.76561$—the vapor rate entering the compressor. In turn, according to the terminology of Fig. 7.3,

$$-W_{comp} = 900 \frac{V_4}{T_4} \left(\frac{-0.5}{900} \right)^{-1} \left\{ \exp\left[\frac{-0.5}{900}(910 - 580) \right] - 1 \right\},$$

where $V_4/T_4 = 0.76561/580$. Therefore, per unit of $V = V_1 = V_2$,

$$-W_{comp} = 358.2 \text{ Btu/lb}.$$

It is assumed that partial condensation occurs between point 3 and point 4, that is, if $T_3 = T_4$.

The heat requirement to return the condensate to the saturated condition during compression can be calculated from

$$\int_{T_4}^{T_1} (C_p)_L (T - T_4) \frac{dV}{dT} \, dT = \text{EXCON},$$

where dV/dT can be determined from the relationship between V and T, and where $(C_p)_L$ is the heat capacity of the condensate. It is sufficient, however, to use an average or mean value for the temperature difference. Thus for all practical purposes, EXCON calculates to

$$(1.0) \frac{450 - 120}{2} (1 - 0.76561) = 38.7 \text{ Btu/lb}.$$

This is the requirement per pound of $V = V_1 = V_2$. A heat capacity of unity in British thermal units per pound per degree Fahrenheit is assumed for the condensate.

For the theoretical cycle efficiency, therefore,

$$\text{Eff}_{SatVap} = \frac{740 - 358.2}{575 - 38.7} = 62.2\%.$$

The theoretical Rankine efficiency is

$$\text{Eff}_{Rankine} = 1 / \left(1 + \frac{900/0.5}{1600 - 120} \right) = 45.1\%.$$

A Carnot efficiency could be approximated by

$$\text{Eff}_{Carnot} = \frac{1600 - 120}{1600 + 460} = 71.8\%.$$

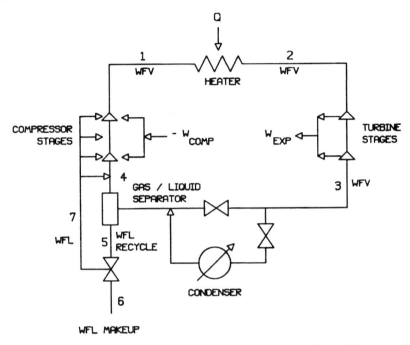

Fig. 7.4. Flow diagram for the saturated vapor cycle with internal evaporative cooling during compression.

The high calculated efficiencies are due to the high value for T_2. The foregoing figures may be adjusted for expansion and/or compression efficiency losses. There is also the serious matter for the initial compression of large actual volumes of gas starting at a low pressure, in this instance, 1.7 psia.

7.4. FLOW DIAGRAM

A flow diagram is shown in Fig. 7.4 for the more involved embodiment, where condensate is produced and reinjected during compression. Such could be referred to as "internal evaporative cooling."

Compressed working fluid vapor (WFV), denoted as stream 1—and which will be at or near the saturated vapor condition—passes through a heater with the addition of heat Q. The vapor is heated to the superheated condition represented by stream 2. Stream 2 then passes through a turbine or turbine stages, which deliver work W_{exp} by gaseous expansion. The exhaust gases (3), which may or may not contain liquids, are sent either directly to a gas/liquid separator or are passed through a condenser for partial condensation or further condensation of working fluid liquid (WFL). The gas or vapor stream (4) leaving the gas/liquid separator flows to the compressor or compressor stages. The bottoms liquid (5), denoted as WFL recycle, may or may not be mixed with WFL makeup (6) to form stream 7, which is injected into or ahead of the compressor or compressor stages. Alternately, WFL makeup (6) may be substituted in its entirety for the WFL recycle. Compression of stream 4 plus the WFL addition (7) to constitute stream 1 completes or closes the cycle.

The work of compression is designated as $(-W_{comp})$, meaning that W_{comp} will have a negative value; that is, in the conventions customarily used, work done on a system is $(-W)$, whereas work done by a system is $(+W)$.

Another way of looking at it is that the previously described saturated vapor cycle can be viewed as a case where the heat of compression is absorbed internally by vaporizing working fluid liquid or condensate, rather than by discharging the heat, say, to a coolant and then to the surroundings. This advantage is partially offset by the fact that condensate has to be produced, which represents a heat loss to the surroundings.

REFERENCE

1. Hougen, O. H., K. M. Watson, and R. A. Ragatz, *Chemical Process Principles. II. Thermodynamics*, Wiley, New York, 1959.

CHAPTER 8

Heat-Pump Cycles That
Use the Saturation Curve

A heat-pump or refrigeration cycle comprises a device that removes heat from the surroundings at a lower temperature and rejects it at a higher temperature. If the higher temperature rejection is of primary interest, the device is referred to as a heat pump. If heat removal at the lower temperature is of main interest, the concept is then referred to as refrigeration. The same mechanical device may serve both purposes by diverting the input and output, one to the other.

This transfer or diversion of heat may be accomplished by a working fluid (or refrigerant) cycle in which the heat of compression raises the temperature of the working fluid. In turn, the working fluid at the higher temperature rejects heat, and after expansion to a lower temperature, receives heat. In principle, this may be regarded as a reverse or inverse Joule cycle, in that heat is rejected at the higher temperature and added at the lower temperature.

The gaseous expansion can be isenthalpic, that is, adiabatic with no work exchange. Alternately, the gaseous working fluid may be expanded through an expander or turbine, and work recouped.

The heat-pump cycle is augmented by operating in the two-phase region, with evaporation and condensation taking place during the cycle; that is, the expansion leg of the cycle starts out as a single-phase liquid, becomes a two-phase mixture as isenthalpic evaporative expansion proceeds, and ends up a gas or vapor. The gas or vapor phase is recompressed following the expansion, then cooled and condensed in a heat exchanger, and the condensate is expanded once again. Thus latent heat effects are involved during heat rejection and heat absorption. Heat is added to the working fluid during or after the expansion and rejected at the cooler–condenser. The nature of the cycle will therefore depend upon the particular working fluid and upon the temperature and pressure levels involved.

A problem common to these devices is a low level of efficiency; that is, the higher the difference in temperature levels across the device, the lower the efficiency. The efficiency can be conveniently expressed in terms of the coefficient of performance (COP), which varies directly with the unit amount of heat transferred and inversely with the temperature difference.

This chapter is directed at increasing or enhancing the efficiency for these types of devices. A method to accomplish this, which will be applied here, is to use compression along the vapor-pressure curve, as developed in the previous chapter. Inasmuch as two-phase operations are already the norm, saturated-vapor compression could as well be included.

The subject of heat pumps that operate in the two-phase region is essentially that of latent heat recovery; the working fluid itself undergoes latent heat changes. Drying was covered previously in Chapter 5, and another example that of a condenser–reboiler coupled distillation column, is presented here in further detail. This chapter is concluded with a representation for multiple-effect evaporation, which is the baseline method for recapturing latent heat during successive evaporative expansions.

8.1. HEAT-PUMP CYCLES

A conventional heat pump is illustrated schematically in Fig. 8.1(a). The corresponding pressure–temperature cycle is diagrammed in Fig. 8.1(b) with respect to the two-phase liquid–vapor region or vapor-pressure curve (also called the saturation curve). The gas at point 1 is compressed to point 2. Then it is cooled, condensed, and subcooled to point 3, by which heat is transferred to lower-temperature surroundings. The liquid at point 3 is then expanded to point 4. If it stays in the liquid region, there is no appreciable change in temperature; that is, the Joule–Thomson coefficient—the change in temperature with pressure during an isenthalpic type of expansion—is very small for liquids. (It in fact exhibits a slightly negative behavior.)

Alternately, the liquid-phase expansion may take place through a mechanical device such as a Pelton wheel. The liquid at point 4 is then heated, vaporized, and superheated to point 5. Heat is therefore transferred from higher-temperature surroundings to the working fluid. The cycle also may be conducted without subcooling the liquid at point 4; that is, point 4 may lie on the saturation curve or vapor-pressure curve.

The foregoing version is of particular interest where it is desirable to utilize the latent heat from the vaporization of the working fluid, along with the latent heat from the condensation of the working fluid.

Another version is represented in Fig. 8.2. Here, the expansion of the liquid from point 3 to point 4 enters the two-phase region and evaporative cooling occurs. Observe that point 3 can remain in or enter into the two-phase region (touch or coincide with the vapor-pressure curve). The expansion from point 3 to point 4 then will occur entirely in the two-phase region.

Finally, in Fig. 8.3 the cycle remains entirely in the single-phase region. The expansion from point 3 to point 4 is a Joule–Thomson or isenthalpic

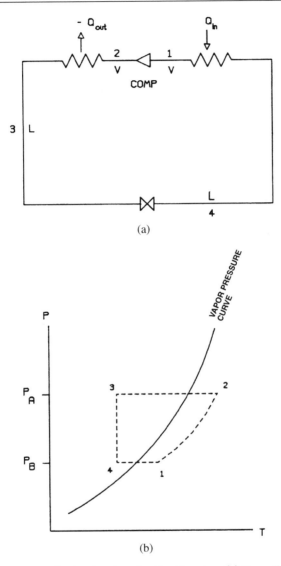

Fig. 8.1. Heat-pump cycle that involves the liquid region. (a) Flow diagram; (b) *P-T* diagram.

expansion of a gas. Alternately, a turbine or other device can be used to extract work during the expansion (e.g., an isentropic expansion).

In all cases, the temperature and pressure levels must be related to the particular working fluid used and its properties, especially the vapor-pressure or saturation curve.

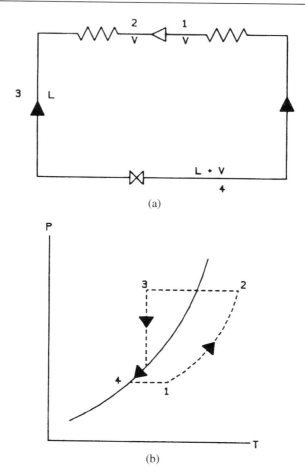

Fig. 8.2. Heat pump in the two-phase region. (a) Flow diagram; (b) *P-T* diagram.

Compression–Expansion Loci. As a matter of note, many or most gases, when compressed, will exhibit a locus of compression that moves away from the vapor-pressure curve. Similarly, the locus of expansion will move toward the vapor-pressure curve. Examples are the lower hydrocarbon gases methane, ethane, and propane, as well as nitrogen, carbon dioxide, hydrogen sulfide, sulfur dioxide, and ammonia, plus steam in particular.

On the other hand, some of the chlorofluorocarbon (CFC) gases may tend to parallel the vapor-pressure curve or else move toward the vapor-pressure curve during compression or move away during expansion. There are no doubt other examples of volatile chlorinated hydrocarbons or

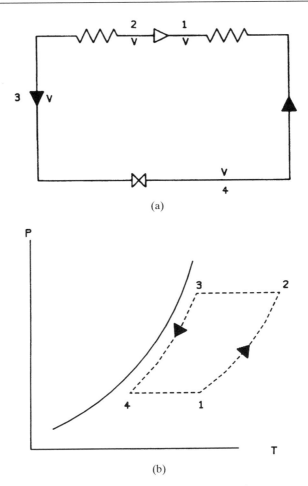

Fig. 8.3. Heat pump in the single-phase gaseous region. (a) Flow diagram; (b) *P-T* diagram.

fluorinated hydrocarbons that behave in this manner in the gaseous or vapor phase.

The gases of the heavier hydrocarbons also exhibit this characteristic of moving toward the curve during compression or away from the curve during expansion. Examples are isobutane and *n*-butane, isopentane and *n*-pentane, and the hexanes and heptanes. Volatile oxygenated compounds such as ethyl ether also may exhibit this characteristic in the gaseous or vapor phase.

Thus there may be many examples of heat-pump working fluids or refrigerants that could serve, but which may have other unsatisfactory problems, such as toxicity or combustibility, or impractical ranges of temperature and pressure.

8.2. HEAT-PUMP CYCLES THAT USE COMPRESSION ALONG THE SATURATED VAPOR CURVE

A heat transfer–compression cycle is described whereby heat may be removed at a lower level and rejected at a higher level at markedly increased efficiencies. This transfer is accomplished by compression along the vapor-phase saturation curve for the working fluid. The injection of working fluid liquid(s) offsets the heat of compression.

Compression along the Vapor-Pressure Curve

By the injection of liquid working fluid or working fluid condensate into the gaseous phase being compressed, it is possible to carry out vapor compression along the two-phase saturation curve or vapor-pressure curve. Vaporization of the condensate partially counteracts the heat of compression—or work of compression. In effect, the temperature or thermal level of the compressor output gases is lowered.

In this fashion, the energy required for compression can be reduced markedly. In other words, the extra thermal energy–pressure–volume work required for vapor superheat thereby is avoided. This is also a feature of the so-called saturated vapor cycle for power generation that previously was examined.

The general features of the heat-pump cycle using compression along the vapor-pressure curve (which also can be called wet compression) are illustrated in Fig. 8.4. The vapor at point 1 is compressed to point 2 while injecting part of the condensate liquids from the condensate at the condition denoted by point 3. The bulk of the condensate (3) is expanded to point 4 and then revaporized to point 1.

Alternately, point 3 can be further subcooled so that point 4 will exist as a subcooled liquid. Under these circumstances there will be a sensible heat absorption as well as a latent heat change when point 4 is heated to point 1.

The working fluid liquid injected during compression may be injected before each compressor stage and/or directly into each stage. The condensate for injection may be taken before or after the expansion. Alternately, makeup working fluid may be added.

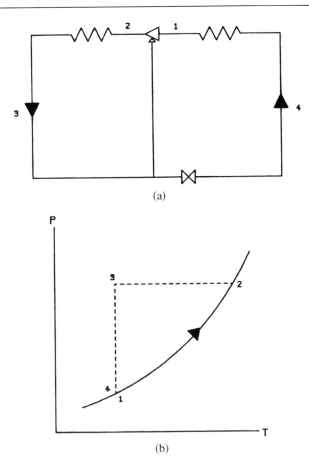

Fig. 8.4. Heat-pump cycle with compression along the vapor-pressure curve. (a) Flow diagram; (b) *P-T* diagram.

Work of Compression along the Vapor-Pressure or Saturation Curve

The relationships for the work of compression along the saturation curve have been derived elsewhere (in Section 6.2). The results are encapsulated and utilized as follows.

The work of compression per unit feed rate is related to the thermal requirement as developed in Section 6.2:

$$-W_{\text{comp}} = C_p(T_2 - T_1) + \Delta H(V_2 - V_1),$$

where $(-W)$ is in heat units and where T_2 and T_1 are on or near the saturated vapor locus, and C_p is the heat capacity of the vapor in consistent units. (Note that in ordinary or isentropic compression, T_2 will be well into the superheated region.)

The vapor mass or molar rates leaving and entering compression are given by V_2 and V_1 (as also developed in Section 6.2). If the working fluid is such that the locus of isentropic compression will follow the vapor-pressure curve, then $(V_2 - V_1) = 0$.

It follows that the ideal or theoretical coefficient of performance (COP), as based on the latent heat $\lambda = \Delta H$ only, will be

$$COP_{theo} = \frac{\lambda}{C_p(T_2 - T_1) + \lambda(V_2 - V_1)}.$$

If the working fluid follows the vapor-pressure curve, as with the Freons, then the second term in the denominator drops out. Here, the COP will depend upon the latent heat as compared to the sensible heat requirement.

The actual COP requires the introduction of the thermoelectrical conversion efficiency and/or the electromechanical efficiency. Representative values are 30 and 80%, respectively. Thus, for example,

$$COP_{act} = COP_{theo}(0.30)(0.80).$$

Although Freons and other working fluids or refrigerants can serve, at first glance it is the very high latent heat of vaporization $\lambda = \Delta H$ of water that may afford an advantage, depending upon the behavior of $(V_2 - V_1)$. However, the temperature levels required also will determine which working fluid is more appropriate; that is, which working fluid can perform satisfactorily in the two-phase vapor–liquid region defined by the temperature levels.

Two-Phase Region

The working fluid may consist of a mixture as well as be composed of a single component. For a mixture, the two-phase vapor–liquid region is defined by an enclosed envelope as shown in Fig. 8.5, along with the cycle. The envelope culminates at the critical point beyond which liquid and vapor can no longer coexist (the supercritical region). Constant percentage or fractional liquid and vapor lines can be drawn inside the envelope. For a pure component, the 100% saturated liquid and vapor lines converge to a

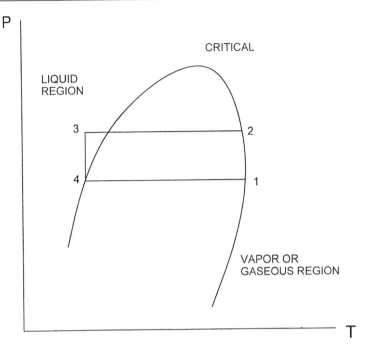

Fig. 8.5. Phase envelope and cycle for a mixture.

single locus that culminates at the critical point and is commonly designated the vapor-pressure curve or saturation curve for the (pure) component or substance.

The principles of operation remain the same for a mixture as for a single component. There will be a change in temperature as vaporization or condensation proceeds at constant pressure, however, or a change in pressure as vaporization or condensation proceeds at constant temperature.

8.3. LATENT HEAT RECOVERY

If the latent heat of condensation of a pure component or mixture is to be recovered and elevated to a higher temperature to supply the latent heat of vaporization, then the working fluid is preferably a two-phase system.

A potential application, at least, lies in condenser–reboiler coupled distillation columns. An example is provided. This example will be followed by a more general treatment that involves the use of compression of the working fluid along the saturation curve as a means to enhance efficiency.

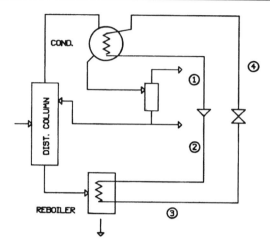

Fig. 8.6. Condenser–reboiler coupled distillation column.

EXAMPLE 8.1

The afore-mentioned cycle is used in conjunction with a distillation column to provide heat for the reboiler and, at the same time, to absorb or receive heat at the condenser, as shown in Fig. 8.6. The condenser temperature is approximately 120°F; the reboiler is 300°F. A figure of 900 Btu/lb will be used for the latent heat of the working fluid and 0.5 Btu/lb-°F will be used for the heat capacity of the working fluid vapor.

Therefore, the ideal or theoretical coefficient of performance is, at best,

$$\text{COP}_{\text{theo}} = \frac{900}{0.5(300 - 120)} = 10.0.$$

The actual COP will be

$$\text{COP}_{\text{act}} = 10.0(0.80)(0.30) = 2.4;$$

that is, the fuel necessary to generate the electricity for the compressor will be only about $1/2.4 = 42\%$ of the heat required for direct heating of the reboiler.

If sensible heat changes are included along with the latent heat change, then the COP will be appreciably higher. (For example, subcooling the condensed working fluid leaving the reboiler, denoted as stream 3, and using this for waste heat or process heat elsewhere.) For instance, in the example, subcooling could add as much as 1.0(300 − 120) = 180 Btu to the latent heat (assuming the heat capacity of the subcooled liquid is about 1.0 Btu/lb-°F).

Process Flow Diagram

A generalized process flow diagram for latent heat recovery is shown in Fig. 8.7. The particulars are further described as follows. Working fluid vapor (WFV), designated as stream 1, is compressed with the addition of working fluid liquid (WFL), denoted as stream 5, at or ahead of the

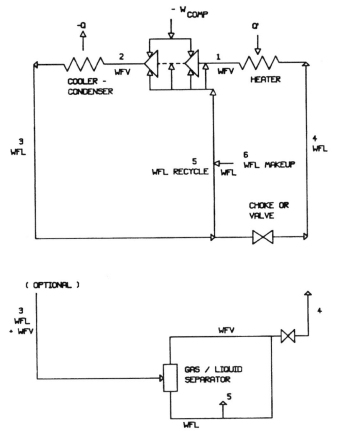

Fig. 8.7. Process flow diagram.

compressor or compressor stages, whereby the path of compression is along or near the saturated vapor locus. The compressed gas or vapor (stream 2) is cooled and condensed or partially condensed in a cooler–condenser, which removes heat $(-Q)$. [In the convention used, $(-Q)$ will be a positive number; that is, if heat is removed from the system, Q is negative.]

If stream 3 leaving the cooler–condenser is entirely a saturated or supercooled liquid (WFL), then part is recycled as stream 5 to the compressor or compressor stages; the bulk and remainder is expanded through a choke or valve to form stream 4. Stream 4 passes through a heater and is vaporized to constitute stream 1 by the addition of heat Q'. This completes the cycle.

Makeup for the working fluid liquid or WFL, designated stream 6, may be added to or used in lieu of stream 5.

If stream 3 is in part liquid (WFL) and part gas or vapor (WFV), then it may be sent to a gas/liquid separator for separation and removal of the liquid (WFL). Part or all of this liquid may be recycled as stream 5 for injection at the compressor or compressor stages. The remainder may be combined with the gas/liquid separator off-gas vapor (WFV) to form stream 4, which proceeds to the heater as before.

8.4. MULTIPLE-EFFECT EVAPORATION

Any consideration of latent heat recovery cycles and systems will have the economies of multiple-effect evaporation as a reference. Such an operation is diagrammed in Fig. 8.8 for a triple-effect system. A notable application is to desalination, for which the previously analyzed cycle also can be used. Stream L_4 represents the liquid feed stream to be vaporized or concentrated, using a steam source V_0. Stream L_1 represents the unvaporized portion or concentrate. Streams C_0, C_1, C_2 denote the successive condensate phases.

The successive evaporation stages 1, 2, 3 will be at decreasing pressures $P_1 > P_2 > P_3$ and decreasing saturation temperatures $T_1 > T_2 > T_3$. Streams L_1, L_2, L_3 will be saturated liquids at the bubble point. Stream L_4 also can be assumed to be a saturated liquid.

Overall and in essence, multiple-effect evaporation can be viewed as a process for removing liquid from dissolved salts or slurries, in which the latent heat of vaporization is recouped or partially recouped. Instead of using a heat pump to achieve higher temperatures, the operating pressures of the sequential evaporators are successively lowered, which progressively

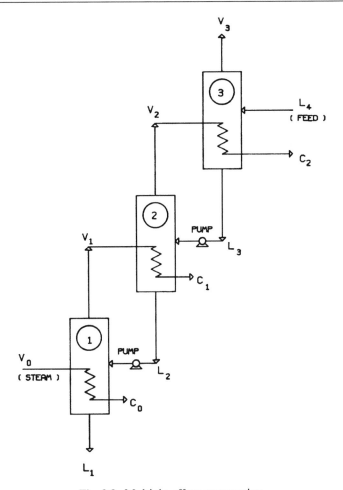

Fig. 8.8. Multiple-effect evaporation.

lowers the evaporation temperatures and thus utilizes waste heat. In other words, lowering the pressure will increase volatility, which provides an alternative to raising the temperature.

Material Balances

The material balances include

$$V_0 = C_0, \qquad V_1 = C_1, \qquad V_2 = C_2,$$
$$L_4 = C_1 + C_2 + L_1 + V_3.$$

Overall Heat Balance

The overall heat balance is

$$V_0 H_{V_0} + L_4 H_{L_4} = C_0 H_{C_0} + C_2 H_{C_1} + C_2 H_{C_2} + L_1 H_{L_1} + V_3 H_{V_3}$$

or

$$V_0 \Delta H_0 = C_1 H_{C_1} + C_2 H_{C_2} + L_1 H_{L_1} + V_3 H_{V_3} - L_4 H_{L_4},$$

where $\Delta H_0 = (H_{V_0}) - (H_{C_0})$ is the latent heat change for the steam used. The quantity $V_0 \Delta H_0$ represents the enthalpic heat input to the system.

If, as a simplification,

$$H_{C_1} \sim H_{C_2} \sim H_{L_1} \sim H_{L_2} \sim H_{L_3} \sim H_{L_4},$$

then

$$V_0 \Delta H_0 \sim (C_1 + C_2 + L_1 - L_4) H_{L_3} + V_3 H_{V_3}$$
$$\sim V_3 (H_{V_3} - H_{L_3})$$
$$\sim V_3 \Delta H_3.$$

This approximation may be used to simplify the coefficient of performance.

Stage Heat Balances

For the third stage,

$$V_2 H_{V_2} - C_2 H_{C_2} = V_3 H_{V_3} + L_3 H_{L_3} - L_4 H_{L_4}.$$

Since $V_2 = C_2$ and $L_3 = L_4 - V_3$,

$$V_2 \Delta H_2 = V_3 (H_{V_3} - H_{L_3}) + L_4 (H_{L_3} - H_{L_4}).$$

Note that if $V_3 \to 0$ and L_4 is saturated, then it would be required that $V_2 = C_2 = 0$. (The condition where $V_3 \to 0$ would represent the bubble point for stream L_3.)

If L_4 is saturated, then for most practical purposes,

$$V_2 \Delta H_2 \sim V_3 \Delta H_3,$$

and if $\Delta H_2 \sim \Delta H_3$, then

$$V_2 \sim V_3.$$

For the second stage,

$$V_1 H_{V_1} - C_2 H_{C_1} = V_2 H_{V_2} + L_2 H_{L_2} - L_3 H_{L_3}$$

or

$$V_1 \Delta H_1 = V_2(H_{V_2} - H_{L_2}) + L_3(H_{L_2} - H_{L_3}).$$

Similarly, if $V_2 \to 0$, then $V_1 = C_1 = 0$.
  If $(H_{L_2}) \sim (H_{L_3})$, then

$$V_1 \Delta H_1 \sim V_2 \Delta H_2,$$

and if $\Delta H_1 \sim \Delta H_2$, then

$$V_1 \sim V_2.$$

  For the first stage,

$$V_0 H_{V_0} - C_0 H_{C_0} = V_1 H_{V_1} + L_1 H_{L_1} - L_2 H_{L_2}$$

or

$$V_0 \Delta H_0 = V_1(H_{V_1} - H_{L_1}) + L_2(H_{L_1} - H_{L_2}),$$

where, if $V_1 \to 0$, then $V_0 = C_0 = 0$.
  Note that if $L_1 \to 0$, this could infer that $V_1 = C_1$ and L_2 are at the same saturation temperature, and therefore heat transfer could not occur at stage 2.
  From the foregoing derivations it follows that

$$V_1 \sim V_2 \sim V_3$$

or

$$C_1 \sim C_2 \sim V_3.$$

Stream Sizes

Usually, for a triple-effect system, streams V_3 and L_1 will be affixed for a given feed stream L_4. Therefore,

$$C_1 + C_2 = L_4 - L_1 - V_3.$$

The relative sizes of C_1 and C_2 will in turn depend upon the pressure and temperature levels and upon the heat transfer surface areas.
  However, as an approximation, as previously determined,

$$C_1 \sim C_2 \sim V_3.$$

Accordingly,

$$C_1 \sim C_2 \sim V_3 = \frac{L_4 - L_1}{3};$$

that is, the vapor and/or condensate rates are apportioned from the difference $L_4 - L_1$.

Coefficient of Performance

An overall coefficient of performance can be based on the heat theoretically required to produce the vapor streams V_1, V_2, V_3 from the respective liquid streams. This theoretical value is the sum

$$Q = (V_1 H_{V_1} + L_1 H_{L_1} - L_2 H_{L_2})$$
$$+ (V_2 H_{V_2}) + (L_2 H_{L_2} - L_3 H_{L_3})$$
$$+ (V_3 H_{V_3} + L_3 H_{L_3} - L_4 H_{L_4})$$
$$= V_1 H_{V_1} + L_1 H_{L_1} + V_2 H_{V_2} + V_3 H_{V_3} - L_4 H_{L_4}.$$

In the preceding case, the latent heat of condensation from streams V_1 and V_2 would not be utilized.

Since

$$L_4 = V_1 + V_2 + L_1 + V_3,$$

the foregoing summation reduces to

$$Q = V_1 (H_{V_1} - H_{L_4}) + V_2 (H_{V_2} - H_{L_4})$$
$$+ V_3 (H_{V_3} - H_{L_4}) + L_1 (H_{L_1} - H_{L_4}).$$

The overall coefficient of performance for the three stages then becomes

$$\text{COP}_3 = \frac{Q}{V_0 \Delta H_0}.$$

Similarly, a COP can be developed for any number of stages. For a single stage, the COP will be unity.

If

$$H_{L_4} \sim H_{L_1} \sim H_{L_2} \sim H_{L_3},$$

then the sum Q becomes

$$Q \sim V_1(H_{V_1} - H_{L_1}) + V_2(H_{V_2} - H_{L_2}) + V_3(H_{V_3} - H_{L_3})$$
$$\sim V_1 \Delta H_1 + V_2 \Delta H_2 + V_3 \Delta H_3$$
$$\sim C_1 \Delta H_1 + C_2 \Delta H_2 + V_3 \Delta H_3.$$

An overall coefficient of performance for the three stages can thus be defined by the simplification

$$\text{COP}_3 \sim \frac{V_1 \Delta H_1 + V_2 \Delta H_2 + V_3 \Delta H_3}{V_0 \Delta H_0}$$

or

$$\sim \frac{C_1 \Delta H_1 + C_2 \Delta H_2 + V_3 \Delta H_3}{V_3 \Delta H_3}.$$

If $\Delta H_1 \sim \Delta H_2 \sim \Delta H_3$ and if $C_1 \sim C_2 \sim C_3$, then

$$\text{COP}_3 \sim 3.$$

For two stages only, it will follow that

$$\text{COP}_2 \sim 2,$$

and for a single stage,

$$\text{COP}_1 = 1.$$

The preceding results may be generalized to n stages, although a larger number of stages may not be practical or economically feasible.

The foregoing analysis assumes certain simplifications in order to relate the number of stages to performance. A detailed analysis of multistage evaporation and the effect of liquid-phase concentrations on boiling temperatures and other such criteria will be found, for instance, in reference 1.

REFERENCE

1. Brown, G. G., A. S. Foust, D. L. Katz, R. Schneidewind, R. R. White, W. P. Wood, G. M. Brown, L. E. Brownell, J. J. Martin, G. B. Williams, J. T. Banchero, and J. L. York, *Unit Operations*, Wiley, New York, 1950.

Symbols

a	constant
a_n	constants for multistage compression ($n = 1, 2, 3, \ldots$)
act	actual
atm	pressure units in atmospheres
A	pressure or temperature level A
A	component A
A	air
A	area
ACF	actual cubic feet
APFM	adjustable proportion fluid mixture
b	constant
b_m	constants for multistage expansion ($m = 1, 2, 3, \ldots$)
B	pressure or temperature level B
B	component B
B	body
B_o	orifice or discharge coefficient (adiabatic)
BP	boiling point
c, c'	constants; constants of integration
comp	compression
C	constant
C	function
C	condensate; condensate rate
C_p	mass or molar heat capacity at constant pressure
$\overline{C}_p$	mean heat capacity (for liquid and vapor phases)
C_v	mass or molar heat capacity at constant volume
ΔC_p	difference between heat capacities of products and reactants
C_o	orifice or discharge coefficient (isothermal); coefficient of flow

C_a coefficient of flow for stream a

C_b coefficient of flow for stream b

C_{a+b} coefficient of flow for stream $a + b$

COP   coefficient of performance

D day

D diameter or equivalent diameter

D_p particle diameter or equivalent particle diameter

elev   elevation

exp   expansion; exponent

E energy

E' energy in force–distance units

E_T total energy

E efficiency

Eff   efficiency

E_{comp} fractional efficiency for compression

E_{exp} fractional efficiency for expansion

EXCON   sensible heat change of condensate

f final

f function, e.g., $f = f(T)$ or $f = f(T_c) = f_c(T_c)$, where T_c is the working fluid temperature during compression

f friction factor ($f = 2f' = 4f''$)

f_{comp} $dq_c/dT = f(T_c) = f_{comp}$

f_{exp} $dq_e/dT = f(T_e) = f_{exp}$

f_c $f_c = f_{comp}$

f_e $f_e = f_{exp}$

F function

F force

F feed; (mass or molar) feed rate

Fl   fluid

g function, e.g., $g = g(T)$ or $g = f_e = f_e(T_e) = f_{exp}$, where T_e is the working fluid temperature during expansion

g acceleration of gravity

g_c gravitational constant; conversion factor between the mechanical units of force–distance and mass–distance

g gas phase

G mass velocity, $G = v\rho$

h pressure drop in inches of water

hp   horsepower

hr   hour

H Henry's constant

H enthalpy or heat function

ΔH latent heat or latent heat change; enthalpy difference

ΔH_v	latent heat of vaporization
ΔH_{WF}	latent heat for working fluid
ΔH	heat of reaction or heat of combustion
ΔH_0	standard heat of reaction or heat of combustion
i	ith variable, term, component, item, or entity
i	moles of component I
I	component I (or inert)
j	jth variable, term, component, item, or entity
J_0	mechanical equivalent of heat
k	kth variable, term, component, item, or entity
k	number of variables or components
k	heat capacity ratio, $k = C_p/C_v$
kW	kilowatt
K	equilibrium distribution function, K-value or equilibrium vaporization ratio, e.g., $K = y/x$
K.E.	kinetic energy
$\underline{lw}$	lost work in consistent units
$\overline{lw}$	lost work per unit mass or per unit mass flow rate
lw*	generalized lost work in nonadiabatic behavior
L	liquid phase; (mass or molar) liquid rate
L	length or distance
LHP	latent heat pump
m	number; number of expansion stages ($m = 1, 2, 3, \dots$)
m	mass or moles; rate
m_0	reference mass or moles; mass or molar flow rate
m	slope; exponent; multiplier
M, MW	molecular weight
M	weight (e.g., of moisture)
M	thousand (e.g., MSCF, thousand standard cubic feet)
MM	million (e.g., MMSCF, million standard cubic feet)
n	number; number of compression stages ($n = 1, 2, 3, \dots$)
n	number of moles, number of moles of oxygen per mole of fuel
n_i	number of moles of component i (or I)
n_0	initial number of moles
n_T	total number of moles
N	total number
p	number of phases
pp	partial pressure
psi	pressure units in pounds per square inch
psia	pressure units in pounds per square inch absolute

psf pressure units in pounds per square foot

P function

P product(s); ΣP is the totality of products; W_P is the total mass flow rate of the products

P pressure

P' pressure in force–distance units

ΔP pressure-drop

P_0 initial pressure

P_1, P_2, P_3, P_4 pressure at corresponding points

P_a, P_b pressure levels

P_A, P_B pressure levels

P_A air pressure

P_c pressure during compression

P_e pressure during expansion

P_c pressure at critical velocity

P_c critical pressure

P_r reduced pressure, $P_r = P/P_c$, where P_c is critical pressure

P_{st} stack gas pressure

P.E. potential energy

q heat added in mechanical units or consistent units

$\bar{q}$ heat added per unit mass or unit mole

Δq heat added, expressed as increment

$\Delta \bar{q}$ heat added per unit mass or mole, expressed as increment

q_c heat added during compression, in mechanical units or consistent units ($-q_c$ = heat lost during compression)

q_e heat added during expansion, in mechanical units or consistent units

Q heat added in heat units

Q_c heat added during compression, in heat units ($-Q_c$ = heat lost during compression)

Q_e heat added during expansion, in heat units

Q_{n-1} heat added prior to nth compression stage ($-Q_{n-1}$ = heat lost)

$\bar{Q}_{m-1}$ heat added prior to mth expansion stage

R component R

R reactant(s); ΣR is the totality of reactants; W_R is the total mass flow rate of the reactants

R function

R temperature ratio for variable interstage heat transfer during multistage compression

R gas constant in consistent units or mechanical units

R_0 gas constant in heat units

R condensation point in Rankine cycle

Re   Reynolds number, $Re = Dv\rho/\mu$

ROI   return on investment

s distance

s, S saturation; stack gas

S vaporization point in Rankine cycle

st   stack gases

S component S

S entropy

S^* generalized entropy function for nonadiabatic behavior

S temperature ratio for variable interstage heat transfer during multistage expansion

S weight (e.g., of solids)

S evaporation point in Rankine cycle

SCF   standard cubic feet

t time

t temperature

theo   theoretical

t   ton

T temperature; temperature in degrees absolute

ΔT temperature difference

ΔT_{lm} log mean temperature difference

T' temperature; temperature after compression; temperature during isentropic change

T_0 initial temperature; reference temperature, e.g., at standard conditions

T_1, T_2, T_3, T_4 temperatures at the corresponding points

T_a, T_b, T_c, T_d temperatures at the corresponding points

T_a temperature level; temperature of air or ambient temperature

T_b temperature level; temperature of combustants; temperature of preheated air or preheated air–fuel mixture

T_c temperature of working fluid during compression

T_e temperature of working fluid during expansion

T_g gaseous or vapor temperature ($= T_V$)

T_c cold-side temperature

T_h hot-side temperature

T_c temperature level; (adiabatic) temperature level of combustion products

T_c temperature at critical velocity

T_c critical temperature

T_r reduced temperature, $T_r = T/T_c$, where T_c is critical temperature

T_n exit temperature from nth compressor stage

T'_{n-1} inlet temperature to nth compressor stage

$\overline{T}_m$ exit temperature from mth expander stage

$\overline{T}'_{m-1}$ inlet temperature to mth expander stage

T_A, T_B temperature levels

T_F feed temperature

T_L liquid temperature

T_V vapor temperature $(= T_g)$

T_C condensate temperature

T_D dried-product temperature

T_P phase-separation temperature

T_S or T_s saturation temperature

T_S stack-gas temperature before heat regeneration

T_{Sf} final stack-gas temperature

T_{st} stack-gas temperature

U internal energy

U overall heat transfer coefficient

v velocity

v_c critical velocity

VP   vapor pressure

v, V vapor phase

V vapor phase; vapor rate (mass or molar)

V volume or specific volume

$\overline{V}$ specific volume, $\overline{V} = 1/\rho$, mass or molar basis

V_c volume at critical velocity

V vapor; (mass or molar) vapor rate

V_g or V_V volume of gaseous or vapor phase

V_L volume of liquid phase

w work done by system in mechanical units or consistent units ($-w$ is the work done on the system)

$\overline{w}$ work done per unit mass or mole

Δw work done, expressed as increment

$\Delta \overline{w}$ work done per unit mass or mole, expressed as increment

w_c work done by compression ($-w_c$ = work required)

w_{comp} work done by compression ($-w_{comp}$ = work required)

w_e work done by expansion

w_{exp} work done by expansion

W work done in heat units

W' work done in heat units

$-\Delta W_p$ parasitic requirement for Rankine cycle

$-W_{pump}$ pumping requirement for Rankine cycle

W_n work done by nth compressor stage ($-W_n$ = work required)

$\overline{W}_m$ work done by mth expander stage

W flow rate, mass or molar

W_a, W_b, W_c, W_d flow rate at the corresponding points

W_a flow rate of stream a

W_b flow rate of stream b

W_{a+b} flow rate of stream $a + b$

W_F flow rate of feed stream; discharge stream; mass flow rate of fuel

W_L flow rate of stream L

W_V flow rate of stream V

W_A flow rate of combustive air (stream A); flow rate of recycle stream

W_B flow rate of combustion products (stream B)

W_C condensate rate

W_D dried-solids rate

W_P flow rate for totality of products, $W_P = \Sigma P$

W_R flow rate for totality of reactants, $W_R = \Sigma R$

W_{WF} working fluid flow rate

WF   working fluid

WFL   working fluid existing as liquid

WFV   working fluid existing as vapor

x x-coordinate

x distance

x weight fraction; moisture content

x_i mole fraction (or mass fraction) of component i (e.g., in liquid phase)

x_j mole fraction (or mass fraction) of component j (e.g., in liquid phase)

y y-coordinate

y distance

y_i mole fraction (or mass fraction) of component i (e.g., in vapor phase)

y_j mole fraction (or mass fraction) of component j (e.g., in vapor phase)

z z-coordinate

z distance; elevation

z_{elev} elevation

z compressibility factor

$\overline{z}$ mean or average compressibility factor

z_A compressibility factor for air

z_{st} compressibility factor for stack gases

z_i mole fraction (or mass fraction) of component i

z_j mole fraction (or mass fraction) of component j

Greek Letters

α temperature ratio for adiabatic isentropic compression

α relative volatility

β temperature ratio for adiabatic isentropic expansion

ε pointwise efficiency

$\bar{\varepsilon}$ mean value of pointwise efficiency

ε_m mean value of pointwise efficiency

η Gay–Lussac coefficient, dT/dV

λ integrating factor

λ latent heat of evaporation

λ or λ_T isothermal Gay–Lussac coefficient, $\lambda = -C_v\eta$

μ Joule–Thomson coefficient, dT/dP

μ viscosity

ρ density

ϕ or ϕ_T isothermal Joule–Thomson coefficient, $\phi = -C_p\mu$

ω intrinsic energy in mechanical units or consistent units

ω^* generalized irreversibility in mechanical units, or consistent units, for nonadiabatic behavior

Ω intrinsic energy in heat units

Ω^* generalized irreversibility in heat units for nonadiabatic behavior

Index

345